W9-ADN-628

HOLOMORPHIC MAPS AND INVARIANT DISTANCES

NORTH-HOLLAND
MATHEMATICS STUDIES **40**

Notas de Matemática (69)

Editor: Leopoldo Nachbin

Universidade Federal do Rio de Janeiro
and University of Rochester

Holomorphic Maps and Invariant Distances

TULLIO FRANZONI
Scuola Normale Superiore,
Pisa, Italy

and

EDOARDO VESENTINI
Scuola Normale Superiore,
Pisa, Italy
and
University of Maryland,
College Park, Maryland, U.S.A.

1980

NORTH-HOLLAND PUBLISHING COMPANY – AMSTERDAM • NEW YORK • OXFORD

© *North-Holland Publishing Company, 1980*

All rights reserved. No part of this publication may be reproduced, stored in a retrieval system, or transmitted, in any form or by any means, electronic, mechanical, photocopying, recording or otherwise, without the prior permission of the copyright owner.

ISBN: 0 444 85436 3

Publishers:

NORTH-HOLLAND PUBLISHING COMPANY
AMSTERDAM • NEW YORK • OXFORD

Sole distributors for the U.S.A. and Canada:

ELSEVIER NORTH-HOLLAND, INC.
52 VANDERBILT AVENUE, NEW YORK, N.Y. 10017

Library of Congress Cataloging in Publication Data

Franzoni, Tullio, 1944-
Holomorphic maps and invariant distances.

(North-Holland mathematics studies ; 40)
(Notas de matemática ; 69)
"These notes originate from lectures given at the University of Maryland and at the Scuola normale superiore, Pisa, in the past two years."
Bibliography: p.
Includes index.
1. Banach spaces. 2. Holomorphic maps. 3. Domains of holomorphy. 4. Distance geometry. I. Vesentini, Edoardo, joint author. II. Title. III. Series. IV. Series: Notas de matemática ; 69.
QA322.2.F72 515.7'32 79-28382
ISBN 0-444-85436-3

PRINTED IN THE NETHERLANDS

Math sep.

PREFACE

These notes originate from lectures given at the University of Maryland and at the Scuola Normale Superiore, Pisa, in the past two years.

The main purpose of the lectures was to give a first systematic, self contained account of the theory of Carathéodory's and Kobayashi's distances and differential metrics on domains in complex Banach spaces.[†)] A complete description of the behaviour of these metrics in concrete cases follows from 'Schwartz lemmas', and these depend on maximum principles. Thus, after two chapters devoted to the theory of holomorphic maps in Banach spaces, a third chapter is concerned with the investigation of various maximum principles and some of their consequences.

The theory of invariant pseudodistances and invariant pseudometrics is developed in the fourth and fifth chapter with special emphasis on the infinite dimensional case, referring to Kobayashi's monograph for finite dimensional domains. The sixth chapter is devoted to the relevant example of the unit ball of a complex Hilbert space. Here the semi-group of all holomorphic isometries for the Carathéodory (= Kobayashi) distance is described, together with the subgroup consisting of all holomorphic automorphisms.

[†)] Only after the final draft of these notes had been completed, the authors became aware of an excellent expository article by L.A. Harris (*Schwartz-Pick systems of pseudometrics for domains in normed linear spaces*, in "Advances in holomorphy (Editor: J.A. Barroso)", Notas de Matematica (65), North-Holland, Amsterdam, 1979; 345-406) which overlaps with some of the material covered by the present work.

In order to keep these notes at an elementary level, no attempt is made to go beyond the case of domains in Banach spaces, although some of the results extend with only minor technical adjustments to complex connected Banach manifolds.

The audience of these lectures consisted mainly of graduate students with only a sketchy knowledge of linear functional analysis and of the differential geometry of the Poincaré metric. Some of the background material on these topics appears in two appendices.

It is a pleasure for the authors to thank Philippe Paclet, Graziano Gentili, Laura Geatti, Michael O'Connor, Edoardo Ballico for their active participation in the lectures. Special thanks go to Fulvio Ricci who contributed significant examples and whose constructive criticism was often instrumental in keeping the lectures in their tracks.

It is a pleasant duty to thank Leopoldo Nachbin for having accepted this book in the "Notas de Matematica". Our gratitude goes also to Kirsti Nicotra for her excellent preparation of these notes.

San Martino di Sarentino, August 1979

T. Franzoni

E. Vesentini

TABLE OF CONTENTS

CHAPTER I

POLYNOMIALS AND POWER SERIES

This chapter is devoted to the elementary theory of vector valued polynomials and power series. The main reference will be Nachbin [1] .

As usual, $\mathbb{Z}$ and $\mathbb{Q}$, $\mathbb{R}$, $\mathbb{C}$ will denote the sets of all integers, and of all rational, real, complex numbers. The symbol $\overline{\mathbb{R}}$ will indicate the extended real line: $\overline{\mathbb{R}} = \mathbb{R} \cup \{-\infty, -\infty\}$, and $\mathbb{R}_+$, $\mathbb{R}_-$, $\mathbb{R}_+^\star$, $\mathbb{R}_-^\star$, $\mathbb{N}^\star$ will be defined by $\mathbb{R}_+ = \{t \in \mathbb{R}: t \geqslant 0\}$, $\mathbb{R}_- = -R_+ = \{t \in \mathbb{R}: t \leqslant 0\}$, $\mathbb{R}_+^\star = \mathbb{R}_+ \setminus \{0\}$, $\mathbb{R}_-^\star = \mathbb{R}_- \setminus \{0\}$, $\mathbb{N}^\star = \mathbb{N} \setminus \{0\}$.

In a space X endowed with a topology τ, the interior part, the closure, the boundary of a subset $T \subset X$ will be denoted by $\overset{\circ}{T}$, $\overline{T}$, ∂T, or - more pedantically - by $\overset{\circ}{T}{}^\tau$, $\overline{T}^\tau$, ∂T^τ. If X is a metric space with distance d, for $x \in X$, $r > 0$, $B(x,r)$ or $B_d(x,r)$ will be the open ball with center x and radius r; Δ and $\Delta^\star$ will be, respectively, the open unit disc in $\mathbb{C}$ and the punctured unit disc $\Delta^\star = \Delta \setminus \{0\}$. The vector spaces $\mathbb{C}^n$ and $\mathbb{R}^n$ will always be endowed with their standard topologies.

§ 1. Multilinear maps and polynomials.

Let E and F be complex normed spaces. For any integer $q \geqslant 1$ we denote by E^q the product space $E \times E \times \ldots \times E$ (q times) with the norm defined by

(I.1.1) $$\|(x_1, x_2, \ldots, x_q)\| = \max(\|x_1\|, \|x_2\|, \ldots, \|x_q\|).$$

Let $L^q(E,F)$ be the space of all continuous q-linear maps $E^q \to F$. It is a complex vector space with respect to point-wise vector operations; it is also a normed space with respect to the norm

(I.1.2) $$\|A\| = \sup\{\|A(x_1,\dots,x_q)\|: x_1,\dots,x_q \in E, \|x_i\| \leq 1\}$$

$(A \in L^q(E,F))$.

The space $L^q(E,F)$ is a Banach space if F is complete. For $q=1$, $L^1(E,F)$ is the normed space of all continuous linear mappings from E to F.

Let $L^q_s(E,F)$ be the vector subspace of $L^q(E,F)$ consisting of all $A \in L^q(E,F)$ which are symmetric, i.e. such that

$$A(x_{\pi(1)},\dots,x_{\pi(q)}) = A(x_1,\dots,x_q)$$

for all $x_1,\dots,x_q$ in E and for every permutation π of $\{1,\dots,q\}$. Clearly $L^q_s(E,F)$ is a closed subspace of $L^q(E,F)$, and therefore is a Banach space if F is complete.

Given any q-linear map $A: E^q \to F$, let A_s be the operator defined by

(I.1.3) $$A_s(x_1,\dots,x_q) = \frac{1}{q!}\sum_{\pi} A(x_{\pi(1)},\dots,x_{\pi(q)}),$$

where the summation is taken over all permutations π of $\{1,\dots,q\}$. Clearly A_s is symmetric. If $A \in L^q(E,F)$, then $A_s \in L^q_s(E,F)$, and

$$\|A_s\| \leq \|A\|.$$

If A is symmetric, then $A_s = A$. Hence the mapping $A \mapsto A_s$ defined by (I.1.3) is a continuous projection of $L^q(E,F)$ onto $L^q_s(E,F)$.

If $q=1$, then $L^1_s(E,F) = L^1(E,F) = L(E,F)$.

For $q=0$ we set

$$L^0(E,F) = L^0_s(E,F) = F.$$

Remark 1. If E is not complete, let $\widetilde{E}$ be the completion of E. Then $\widetilde{E}^q$ is the completion of E^q for the norm (I.1.1). If F is complete, every $A \in L^q(E,F)$ has a unique extension $\widetilde{A} \in L^q(\widetilde{E},F)$, and $\|\widetilde{A}\| = \|A\|$.
If $A \in L^q(E,F)$, then $(\widetilde{A_s}) = (\widetilde{A})_s$, and so, if $A \in L^q_s(E,F)$, then $\widetilde{A} \in L^q_s(\widetilde{E},F)$.

Let A be a q-linear map of E^q into F. We denote by $\hat{A}$ the map of E into F defined by the restriction of A to the diagonal:

$$\hat{A}(x) = A(x,\ldots,x).$$

Such an $\hat{A}$ will be called a q-homogeneous polynomial from E to F. Thus, a map $P: E \to F$ is a q-*homogeneous polynomial* if there is a q-linear map $A: E^q \to F$ such that $P = \hat{A}$. Note that $\hat{A} = \hat{A}_s$.

Lemma I.1.1. *(Polarization formula). For any symmetric* q-*linear form* $A: E^q \to F$ *the following identity holds:*

$$A(x_1,\ldots,x_q) = \frac{1}{q!2^q} \sum_{\substack{\varepsilon_1=\pm 1\\ \vdots \\ \varepsilon_q=\pm 1}} \varepsilon_1\cdots\varepsilon_q \hat{A}(\varepsilon_1 x_1+\varepsilon_2 x_2+\ldots+\varepsilon_q x_q)$$

$(x_1,\ldots,x_q \in E)$.

Proof. The above identity is trivial for $q=0,1$. We prove it for $q>1$. Note first that, for any positive integer p,

$$\text{(I.1.4)} \qquad \sum_{\substack{\varepsilon_1=\pm 1\\ \cdots \\ \varepsilon_p=\pm 1}} \varepsilon_1 \cdots \varepsilon_p = (1-1) \cdots (1-1) = 0.$$

The coefficient of $A(x_1,\ldots,x_q)$ on the right hand side is

$$\frac{1}{q!2^q}\sum_{\substack{\varepsilon_1=\pm 1\\ \dots\\ \varepsilon_q=\pm 1}} \varepsilon_1^2 \dots \varepsilon_q^2 q! = \frac{1}{2^q}\underbrace{(1+1)\ \dots\ (1+1)}_{q \text{ times}} = \frac{1}{2^q}\cdot 2^q = 1.$$

All other summands on the right hand side contain terms of type $A(x_{j_1},\dots,x_{j_q})$, where $1 \leqslant j_1 \leqslant \dots \leqslant j_q \leqslant q$, but $j_1,\dots,j_q$ are not all distinct. The coefficient of such an $A(x_{j_1},\dots,x_{j_q})$ is $\frac{1}{q!2^q}\sum_{\varepsilon_i=\pm 1} \varepsilon_1 \dots \varepsilon_p (q-p)!$, where p is the number of distinct indices in $(j_1,\dots,j_q)$.

Thus, by (I.1.4), the coefficient vanishes and the lemma is proven. Q.E.D.

Exercise. Following the above arguments prove the following generalization of the polarization formula:

$$A(y,\dots,y,\ z_1,\dots,z_r) = \frac{1}{q-r+1}\cdot\frac{(q-r)!}{2^r q!}\sum_{\varepsilon_i,\eta} \varepsilon_1 \dots$$
$$\dots \varepsilon_r \eta \hat{A}(\varepsilon_1 z_1 + \dots + \varepsilon_r z_r + \eta y),$$

where $A\colon E^q \to F$ is a symmetric q-linear form, $1 \leqslant r \leqslant q$, each of the ε_i is ± 1, and η varies on the $(q-r+1)$-th roots of 1.

The set of all q-homogeneous polynomials is a complex vector space with respect to pointwise composition. Let $P^q(E,F)$ be the vector subspace consisting of all continuous q-homogeneous polynomials. For the extreme values $q=0,1$, we have $P^0(E,F) = F, P^1(E,F) = L(E,F)$.

If $A \in L^q(E,F)$, i.e. if A is bounded in a neighborhood of $0 \in E^q$, then $\hat{A}$ is continuous. Viceversa, if P is a continuous q-homogeneous polynomial, then P is bounded on a neighborhood of 0 in E. Lemma I.1.1. implies that the symmetric q-linear operator defined by the polarization formula is bounded on a neighborhood of 0 in E^q. Thus we have

<u>Lemma I.1.2</u>. *A q-homogeneous polynomial* P *from* E *to* F *is continuous if, and only if, there is a q-linear map* $A \in L^q(E,F)$ *such that* $P = \hat{A}$.

For any $P \in P^q(E,F)$ we define the norm $\|P\|$ of P, setting

$$\|P\| = \sup\{\|P(x)\| : x \in E,\ \|x\| \leqslant 1\}; \tag{I.1.5}$$

hence

$$\|P(x)\| \leqslant \|P\| \cdot \|x\|^q \qquad \text{for all } x \in E. \tag{I.1.6}$$

The space $P^q(E,F)$ becomes then a normed space, which is complete if F is complete.

<u>Remark 2</u>. If F is complete but E is not complete, denoting again by $\tilde{E}$ the completion of E, Remark 1 and Lemma I.1.2 imply that every $P \in P^q(E,F)$ has a unique extension $\tilde{P} \in P^q(\tilde{E}\ F)$, and $\|\tilde{P}\| = \|P\|$.

<u>Proposition I.1.3</u>. *The mapping* $A \mapsto \hat{A}$ *is a continuous isomorphism of* $L_s^q(E,F)$ *onto* $P^q(E,F)$. *Furthermore,*

$$\|\hat{A}\| \leqslant \|A\| \leqslant \frac{q^q}{q!}\|\hat{A}\|. \tag{I.1.7}$$

<u>Proof</u>. The first part of the statement follows from the inequalities (I.1.7). The first of these is a consequence of (I.1.2) and (I.1.5). To prove the second part, put $\|x_1\| \leqslant 1, \dots, \|x_q\| \leqslant 1$ in the polarization formula. Since, by

(I.1.6), $\|\hat{A}(\varepsilon_1 x_1+\ldots+\varepsilon_q x_q)\| \leqslant \|\hat{A}\| \ \|\varepsilon_1 x_1+\ldots+\varepsilon_q x_q\|^q \leqslant$
$\leqslant \|\hat{A}\| \ (\|x_1\|+\ldots+\|x_q\|)^q \leqslant \|\hat{A}\| \cdot q^q,$

Lemma I.1.1 yields

$$\|A(x_1, \ldots x_q)\| \leqslant \frac{1}{q!\,2^q} \|\hat{A}\| q^q \cdot 2^q = \frac{q^q}{q!} \|\hat{A}\|$$

for all $x_1,\ldots,x_q$ such that $\|x_1\|, \ldots, \|x_q\| \leqslant 1$. Q.E.D.

Examples. 1. Let $E = \mathbb{C}^n$ (with the natural topology, defined by any norm), and let $F = \mathbb{C}$ (natural topology). Any ordinary homogeneous polynomial function of degree q on E is a continuous q-homogeneous polynomial, and viceversa.

2. Set in the above example $n=q$, and consider on $E = \mathbb{C}^q$ the norm

$$\|x\| = |\zeta^1|+\ldots+|\zeta^q| \qquad (x = (\zeta^1, \ldots, \zeta^q)).$$

Let $P \in P^q(E,\mathbb{C})$ be defined by

$$P(x) = \zeta^1 \ldots \zeta^q.$$

Then $P = \hat{A}$, for $A \in L^q_s(E,F)$ given by

$$A(x_1, \ldots, x_q) = \frac{1}{q!} \Sigma\, \zeta^1_{j_1} \ldots \zeta^q_{j_q}, \tag{I.1.8}$$

where $x_j = (\zeta^1_j, \ldots, \zeta^q_j)$ $(j=1, \ldots, q)$ and where the summation is extended over all permutations $(j_1, \ldots, j_q)$ of $(1, \ldots, q)$.

Then

$$\|P\| = \frac{1}{q^q}, \tag{I.1.9}$$

while

$$\|A\| = \frac{1}{q!}. \tag{I.1.10}$$

This shows that the constant $\frac{q^q}{q!}$ appearing in (I.1.7) is the best possible.

3. Set in Example 1 $n=1$, and let $P \in \mathcal{P}^q(\mathbb{C},\mathbb{C})$ be defined by $P(\zeta) = \zeta^q$.

Then $P = \hat{A}$, where $A \in \mathcal{L}^q(\mathbb{C},\mathbb{C})$ is given by $A(\zeta_1, \dots, \zeta_q) = \zeta_1 \dots \zeta_q$.

Thus $\|P\| = \|A\| = 1$. On the other hand

$$\frac{q^q}{q!} = \frac{q^{q-1}}{(q-1)!} > 1 \qquad \text{for} \quad q>1.$$

This example shows that, for particular spaces and for particular values of q, the constant $\frac{q^q}{q!}$ is not the best possible.

4. Let $E = \ell_1(\mathbb{N}^\star) = \{x = (\zeta^\nu)_{\nu=1,2,\dots}: \|x\| = \sum_1^\infty |\zeta^\nu| < +\infty\}$, and let $F = \mathbb{C}$.

For $q \geqslant 1$, let $P \in \mathcal{P}^q(E,\mathbb{C})$ be defined by

$$P(x) = \zeta^1 \dots \zeta^q \qquad (x = (\zeta^1, \dots, \zeta^q, \dots)).$$

Then $P = \hat{A}$, where $A \in \mathcal{L}_s^q(E,\mathbb{C})$ is expressed by (I.1.8).

Also in this case (I.1.9) and (I.1.10) hold.

5. Let E be a pre-Hilbert space, F a normed space, and let $A \in \mathcal{L}_s^2(E,F)$. Then the polarization formula reads

$$A(x_1,x_2) = \frac{1}{2\cdot 2^2} \sum_{\varepsilon_1,\varepsilon_2=\pm 1} \hat{A}(\varepsilon_1 x_1 + \varepsilon_2 x_2).$$

Hence

$$\|A(x_1,x_2)\| \leqslant \frac{1}{2\cdot 2^2} \|\hat{A}\| (2\|x_1+x_2\|^2 + 2\|x_1-x_2\|^2) =$$

$$= \frac{1}{2^2} \|\hat{A}\| (\|x_1+x_2\|^2 + \|x_1-x_2\|^2) = \frac{1}{2^2}\|\hat{A}\| (2\|x_1\|^2 + 2\|x_2\|^2) =$$

$$= \frac{1}{2} \|\hat{A}\| (\|x_1\|^2 + \|x_2\|^2).$$

Thus $\|A\| \leqslant \|\hat{A}\|$, and so the first inequality in (I.1.7) proves that, if E is a pre-Hilbert space, for every $A \in L_s^2(E,F)$ we have

$$\|A\| = \|\hat{A}\|.$$

As it was noticed before, a q-homogeneous polynomial is continuous if, and only if, it is bounded on a neighborhood of the origin.

Definition. Let U be a subset of E. A function $f: U \to F$ is *locally bounded* if every $x \in U$ has a neighborhood on which f is bounded.

Clearly, if f is continuous, then f is locally bounded. We will prove that the converse to this statement holds for homogeneous polynomials.

The arguments preceding Lemma I.1.2 show that a homogeneous polynomial is continuous if, and only if, it is bounded on a neighborhood of 0. If a homogeneous polynomial is bounded on some ball centered at 0, then it is bounded on any ball centered at 0.

Now, let P be a q-homogeneous polynomial from E to F, and let A be a symmetric q-linear map of E^q into F such that $\hat{A} = P$. For $y,z \in E$, the Newton identity holds:

$$\hat{A}(y+z) = \sum_{p=0}^{q} \binom{q}{p} A(\underbrace{y,\dots,y}_{q-p},\underbrace{z,\dots,z}_{p}).$$

Let C_y^{q-p} be the symmetric p-linear map of E^p into F defined by

$$C_y^{q-p}(z_1,\dots,z_p) = A(y,\dots,y,z_1,\dots,z_p).$$

If A is continuous, then

$$\|C_y^{q-p}(z_1,\dots,z_p)\| \leq \|A\| \cdot \|y\|^{q-p} \|z_1\| \dots \|z_p\|,$$

i.e.

$$\|C_y^{q-p}\| \leqslant \|A\| \cdot \|y\|^{q-p}.$$

The Newton identity reads now

$$\text{(I.1.11)} \qquad P(y+z) = \hat{A}(y+z) = \sum_{p=0}^{q} \binom{q}{p} \widehat{C_y^{q-p}}(z),$$

i.e.

$$\text{(I.1.12)} \qquad P(x) = \hat{A}(x) = \hat{A}(y+x-y) = \sum_{p=0}^{q} \binom{q}{p} \widehat{C_y^{q-p}}(x-y).$$

Suppose now that the q-homogeneous polynomial P is bounded on a neighborhood of some point $y \in E$. Thus there is a finite constant $K \geqslant 0$ and some $R > 0$ such that

$$\text{(I.1.13)} \qquad \|P(y+z)\| \leqslant K \quad \text{for all} \quad z \in E \quad \text{with} \quad \|z\| \leqslant R.$$

By the formula of the exercise following Lemma I.1.1:

$$A(y,\dots,y,\ z_1,\dots,z_r) = \frac{1}{(q-r+1)} \cdot \frac{(q-r)!}{2^r q!} \sum_{\varepsilon_i, \eta} \varepsilon_1 \cdots$$
$$\cdots \varepsilon_r \eta \hat{A}(\varepsilon_1 z_1 + \dots + \varepsilon_r z_r + \eta y),$$

where each of the ε_i, $1 \leqslant i \leqslant r$, is ± 1, and η varies on the $(q-r+1)$-th roots of 1.

Hence, by (I.1.13), there exists a finite constant $K_1 \geqslant 0$ such that

$$\|\widehat{C_y^{q-p}}(z)\| \leqslant K_1 \quad \text{whenever} \quad \|z\| \leqslant R \quad (p=0,1,\dots,q),$$

or, equivalently,

$$\|\widehat{C_y^{q-p}}(z)\| \leqslant \frac{K_1}{R^p} \|z\|^p \qquad \text{for all} \quad z \in E \quad (p=0,1,\dots,q).$$

Formula (I.1.12) shows then that $x \mapsto P(x)$ is bounded on any ball with center 0 and finite radius, and therefore P is continuous. That proves

<u>Theorem I.1.4</u>. *A q-homogeneous polynomial* P *from* E *to* F

is continuous if, and only if, there is an open ball in E *on which* P *is bounded.*

Definition. A *(continuous) polynomial* P from E to F is a mapping $P\colon E \to F$ for which there exist (continuous) q-homogeneous polynomials P_q $(q=0,1,\dots,d)$ such that

$$P(x) = P_0 + P_1(x) + \dots + P_d(x)$$

for all $x \in E$.

The homogeneous polynomials P_q are uniquely determined by P, as it follows from

Lemma I.1.5. *If* $P_0+P_1(x)+\dots+P_d(x) = 0$ *for all* $x \in E$, *then* $P_0 = 0$, $P_1 = 0, \dots, P_d = 0$.

Proof. For every $x \in E$, and every $\zeta \in \mathbb{C}$

$$P_0+P_1(\zeta x)+\dots+P_d(\zeta x) = 0,$$

i.e.

$$P_0+\zeta P_1(x)+\dots+\zeta^d P_d(x) = 0,$$

or, equivalently, for $\zeta \neq 0$,

$$\frac{1}{\zeta^d} P_0 + \frac{1}{\zeta^{d-1}} P_1(x) + \dots + P_d(x) = 0.$$

Letting $|\zeta| \to \infty$ we get $P_d(x) = 0$ for all $x \in E$, i.e. $P_d = 0$. Iterating this procedure we show that $P_{d-1} = 0, \dots, P_0 = 0$.

Corollary I.1.6. *For every polynomial* P *there is a unique way of writing*

$$P = P_0 + P_1 + \dots + P_d,$$

where P_q *is a* q-*homogeneous polynomial from* E *to* F, *and* $P_d \neq 0$.

The integer d is called the *degree* of the polynomial P. We will denote by $P(E;F)$ the set of all continuous polynomials from E to F. It is a vector space for pointwise composition, and $P^q(E,F)$ is a vector subspace of $P(E,F)$ for every $q=0,1,2\ldots$

The following theorem improves theorem I.1.4.

Theorem I.1.7. *Let* P *be a polynomial of degree* d*, mapping the Banach space* E *into a normed space* F*. If there exists a non-empty open set* $U \subset E$ *and a residual set* C *in* U *such that* P *is bounded on* C*, then* P *is a continuous polynomial.*[★)]

This theorem follows from

Lemma I.1.8. *The polynomial* P *is continuous if for some* $x_0 \in E$ *there exists an open neighborhood* B *of* 0 *in* E *and a residual set* $C \subset B$ *such that for every sequence* (u_ν)*,* $u_\nu \in C$*, converging to* 0*, the sequence* $(P(x_0+u_\nu))$ *is bounded.*

Proof. 1) By the Newton identity, for every $y \in E$, we can express $P(x_0+y)$ as a sum

$$P(x_0+y) = \sum_{q=0}^{d} R_q(y),$$

where R_q is a q-homogeneous polynomial from E to F. Hence there is no restriction in assuming $x_0=0$.

2) We will prove the lemma by showing that P is bounded on *every* sequence (y_ν) converging to 0 in E.

Let $S = B\setminus C$.

For $m,n = 0,1,\ldots,d$ and $\nu = 1,2,\ldots,$ let

$$S_{\nu,m,n} = \{ \frac{\nu}{m+1} x - \nu n y_\nu : x \in S\}.$$

The set $S_{\nu,m,n}$ is a homotetic image of S. Since S is

★) All basic notions on residual sets, meager set, etc. can be found in Appendix B.

meager in the open set B, then S is meager in E. Thus $S_{\nu,m,n}$ is meager in E, and therefore the union $T = \cup S_{\nu,m,n}$ does not exhaust E.

Let $y_0 \in E \setminus T$.

If

$$x_{m,n,\nu} = (m+1)(ny_\nu + \frac{1}{\nu} y_0) \in S,$$

then

$$y_0 = \frac{\nu}{m+1} x_{m,n,\nu} - n\nu y_\nu \in S_{\nu,m,n} \subset T.$$

Thus

$$x_{m,n,\nu} = (m+1)(ny_\nu + \frac{1}{\nu} y_0) \notin S$$

for $m,n = 0,1,\dots,d$ and $\nu = 1,2,\dots$.

Moreover, $\lim_{\nu\to\infty} (m+1)(ny_\nu + \frac{1}{\nu} y_0) = 0$ uniformly with respect to $m,n = 0,1,\dots,d$.

Thus there exists a constant $K_1 \geqslant 0$ such that

$$\|P(m+1)(ny_\nu + \frac{1}{\nu} y_0))\| \leqslant K_1 \tag{I.1.14}$$

for all $m,n = 0,1,\dots,d$; $\nu = 1,2,\dots$.

3) We will prove now that P is bounded on the sequence (y_ν).

Since

$$P((m+1)(ny_\nu + \frac{1}{\nu} y_0)) = R_0 + (m+1)R_1(ny_\nu + \frac{1}{\nu} y_0) +$$

$$+ (m+1)^2 R_2(ny_\nu + \frac{1}{\nu} y_0) + \dots + (m+1)^d R_d(ny_\nu + \frac{1}{\nu} y_0)$$

for $m = 0,1,\dots,d$, there exist constants $a_{q0},\dots,a_{qd}$ $(q = 0,1,\dots,d)$ such that

$$R_q(ny_\nu + \frac{1}{\nu} y_0) = a_{q0}P(ny_\nu + \frac{1}{\nu} y_0) + a_{q1}P(2(ny_\nu + \frac{1}{\nu} y_0)) +$$

$$+ \dots + a_{qd}P((d+1)(ny_\nu + \frac{1}{\nu} y_0)).$$

Hence, by (I.1.14), there exists a constant $K_2 \geqslant 0$ for which

(I.1.15) $\quad \| R_q(ny_\nu + \frac{1}{\nu} y_0) \| \leqslant K_2$ for all $q=0,1,\dots,d$, $\nu=1,2,\dots$.

Denoting by A_q the q-linear symmetric operator $E^q \to F$ such that $\hat{A}_q = R_q$, the Newton identity yields

(I.1.16)
$$R_q(ny_\nu + \frac{1}{\nu} y_0) = \sum_{p=0}^{q} \binom{q}{p} A_q(\underbrace{\frac{1}{\nu} y_0,\dots,\frac{1}{\nu} y_0}_{q-p}, \underbrace{ny_\nu,\dots,ny_\nu}_{p})$$
$$= \sum_{p=0}^{q-1} \binom{q}{p} \frac{n^p}{q-p} A_q(\underbrace{y_0,\dots,y_0}_{q-p}, \underbrace{y_\nu,\dots,y_\nu}_{p}) + n^q R_q(y_\nu)$$
$(n=0,1,\dots,d)$.

Denoting by $\mathcal{R}_q$ and $\mathcal{A}_q$ the vectors

$$\mathcal{R}_q = {}^t(R_q(\tfrac{1}{\nu}y_0),\ R_q(y_\nu + \tfrac{1}{\nu}y_0),\dots,R_q(qy_\nu + \tfrac{1}{\nu}y_0)),$$
$$\mathcal{A}_q = {}^t(A_q(y_0,\dots,y_0), A_q(y_0,\dots,y_0,y_\nu),\dots,A_q(y_0,y_\nu,\dots,y_\nu), R_q(y_\nu))$$

and by $\mathcal{L}_q$ the matrix

$$\mathcal{L}_q = \left\| \begin{matrix} \frac{1}{\nu^q} & 0 & 0 & 0 & \dots & 0 & 0 \\ \frac{1}{\nu^q} & \frac{q}{\nu^{q-1}} & \frac{\binom{q}{2}}{\nu^{q-2}} & \frac{\binom{q}{3}}{\nu^{q-3}} & \dots & \frac{q}{\nu} & 1 \\ \frac{1}{\nu^q} & 2\,\frac{q}{\nu^{q-1}} & 2^2\,\frac{\binom{q}{2}}{\nu^{q-2}} & 2^3\,\frac{\binom{q}{3}}{\nu^{q-3}} & \dots & 2^{q-1}\,\frac{q}{\nu} & 2^q \\ \frac{1}{\nu^q} & 3\,\frac{q}{\nu^{q-1}} & 3^2\,\frac{\binom{q}{2}}{\nu^{q-2}} & 3^3\,\frac{\binom{q}{3}}{\nu^{q-3}} & \dots & 3^{q-1}\,\frac{q}{\nu} & 3^q \\ \dots & \dots & \dots & \dots & \dots & \dots & \dots \\ \frac{1}{\nu^q} & q\,\frac{q}{\nu^{q-1}} & q^2\,\frac{\binom{q}{2}}{\nu^{q-2}} & q^3\,\frac{\binom{q}{3}}{\nu^{q-3}} & \dots & q^{q-1}\,\frac{q}{\nu} & q^q \end{matrix} \right\|$$

(I.1.16) can be written

$$R_q = L_q A_q \,.$$

Now

$$\det L_q = c_q \left(\nu^{\frac{q(q-1)}{2}} \right)^{-1}$$

where

$$c_q = 2\cdot 3\cdot \ldots \cdot q\cdot q \binom{q}{2} \ldots \binom{q}{q-1} \det \begin{Vmatrix} 1 & 1 & \ldots & 1 \\ 1 & 2 & \ldots & 2^{q-1} \\ \ldots & \ldots & \ldots & \ldots \\ 1 & q & \ldots & q^{q-1} \end{Vmatrix} \neq 0 .$$

On the other hand, the determinant of the matrix $L_{q,p}$, obtained by suppressing the p-th row $(p=0,1,\ldots,q)$ and the last column in L_q is $\det L_{q,p} = \dfrac{c_{q,p}}{\nu^{q+q-1+\ldots+1}} = \dfrac{c_{q,p}}{\nu^{\frac{(q+1)q}{2}}}$, where $c_{q,p}$ is a real constant.

The Cramer rule yields

$$R_q(y_\nu) = \frac{(-1)^q}{\det L_q} \{\det L_{q,0}\, R_q(\tfrac{1}{\nu} y_0) - \det L_{q,1}\, R_q(y_\nu + \tfrac{1}{\nu} y_0) + \ldots$$
$$\ldots + (-1)^q \det L_{q,q}\, R_q(q y_\nu + \tfrac{1}{\nu} y_0)\} =$$

$$= \frac{(-1)^q}{c_q} \sum_{p=0}^{q} (-1)^p c_{q,p}\, R_q(n y_\nu + \tfrac{1}{\nu} y_0) .$$

Taking into account (I.1.15), we see that there exists a constant $K_3 > 0$ such that

$$\| R_q(y_\nu) \| \leq K_3 \quad \text{for } q=0,1,\ldots,d;\ \nu=1,2,\ldots . \qquad \text{Q.E.D.}$$

As a consequence of the above theorem and of the Osgood theorem we have

<u>Theorem I.1.9</u>. *Let* $(P_q^{(\nu)})$, $\nu=1,2,\ldots$ *be a sequence of continuous* q-*homogeneous polynomials* $P_q^{(\nu)} \in P^q(E,F)$, *converging at every* $x \in E$. *Let* $P: E \to F$ *be the limit function:*

$$P(x) = \lim_{\nu\to\infty} P_q^{(\nu)}(x).$$

Then P *is (either identically zero or) a* q*-homogeneous continuous polynomial.*

Note. It is well known that every continuous linear operator A from the Banach space E to the Banach space F is also continuous with respect to the weak topologies.

Namely, if (x_α) is a net converging weakly to $x \in E$, and λ is any continuous linear form on F, we have

$$\langle Ax_\alpha, \lambda\rangle = \langle x_\alpha, A^\star \lambda\rangle \to \langle x, A^\star\lambda\rangle = \langle Ax, \lambda\rangle ,$$

i.e. (Ax_α) converges to x with respect to the weak topology of F.

This property does not hold for polynomials of higher degree. Let us consider, for example, the Hilbert space $\ell_2(\mathbb{N})$, and the following homogeneous polynomial of degree 2 from $\ell_2(\mathbb{N})$ to $\mathbb{C}$:

$$P(z) = \sum_{i=0}^{\infty} \zeta_i^2 \qquad \text{for} \quad z = (\zeta_0, \zeta_1, \dots, \zeta_n, \dots).$$

Obviously, P is (strongly) continuous (and $\|P\|=1$), but P is not weakly continuous.

In fact, let (e_n) $n=0,1,2,\dots$ be the canonical orthonormal basis of $\ell_2(\mathbb{N})$.

Then, (e_n) converges weakly to 0, but $P(e_n) = 1$ for all $n \in \mathbb{N}$.

§ 2. Convergent power series.

By *power series* from E to F we mean a formal expression $\sum_{q=0}^{+\infty} P_q$, where $P_q \in P^q(E,F)$.

Let $B(0,r) = \{x \in E: \|x\| < r\}$ be the open ball with center 0 and radius $r > 0$ in E.

Definition. The *radius of convergence* of the power series $\sum_{q=0}^{\infty} P_q$ is the largest R $(0 \leqslant R \leqslant +\infty)$ such that the series $\sum_{q=0}^{\infty} P_q(x)$ is *uniformly convergent* for $x \in \overline{B(0,r)}$ for any r with $0 \leqslant r < R$.

If $R > 0$, the power series $\sum_{q=0}^{+\infty} P_q$ is said to be *convergent*.

Let $f(x) = \sum_{q=0}^{+\infty} P_q(x) \qquad (x \in B(0,R))$.

Lemma I.2.1. *(Cauchy-Hadamard formula). The radius of convergence* R *satisfies the inequality*

$$\frac{1}{R} \geqslant \limsup_{q \to \infty} \|P_q\|^{1/q} ,$$

equality holding if F *is complete.*

Proof. The proof is similar to that of the classical Cauchy-Hadamard, and will be reproduced here only for the sake of completeness. We show first that

$$\text{(I.2.1)} \qquad \frac{1}{R} \geqslant \limsup \|P_q\|^{1/q} .$$

If the power series converges uniformly on some $\overline{B(0,r)}$, there is some integer M such that

$$\|f(x) - \sum_{q=0}^{m} P_q(x)\| \leqslant 1$$

for all $x \in \overline{B(0,r)}$ and all $m \geqslant M$.

Since

$$\left| \|f(x) - \sum_{q=0}^{m-1} P_q(x)\| - \|P_m(x)\| \right| \leqslant \|f(x) - \sum_{q=0}^{m} P_q(x)\| \leqslant 1,$$

for all $x \in \overline{B(0,r)}$ and all $m \geqslant M$, then $\|P_m(x)\| \leqslant 2$ for all $x \in \overline{B(0,r)}$ and all $m > M$.

If $\|x\| \leq 1$, then $rx \in \overline{B(0,r)}$, and therefore

$$r^m \|P_m(x)\| \leq 2$$

for all $x \in \overline{B(0,1)}$ and all $m > M$.

Thus $r^m \|P_m\| \leq 2$ for all $m > M$, and therefore

$$r \limsup \|P_m\|^{1/m} \leq 1,$$

i.e. $\frac{1}{r} \geq \limsup \|P_m\|^{1/m}$ for any r such that $0 \leq r < R$. That proves (I.2.1).

Let F be a Banach space.

The inequality opposite to (I.2.1) holds obviously if $\limsup \|P_q\|^{1/q} = \infty$.

Suppose $\limsup \|P_q\|^{1/q} < +\infty$, and choose r such that

$$\text{(I.2.2)} \qquad 0 \leq r < \frac{1}{\limsup \|P_q\|^{1/q}} .$$

For any $t \in (r, \frac{1}{\limsup \|P_q\|^{1/q}})$, there exists some index q_0 such that $\|P_q\|^{1/q} < \frac{1}{t}$ for all $q \geq q_0$, i.e.

$$\|P_q\| < \frac{1}{t^q} \qquad \text{for all } q \geq q_0.$$

Thus, for any $x \in \overline{B(0,r)}$ and all $q \geq q_0$, we have, by (I.1.6),

$$\|P_q(x)\| \leq \|P_q\| \|x\|^q \leq \frac{\|x\|^q}{t^q} \leq (\frac{r}{t})^q ,$$

and therefore

$$\sum_{q=q_0}^{+\infty} \|P_q(x)\| \leq (\frac{r}{t})^{q_0} (1 + \frac{r}{t} + \dots + \frac{r^n}{t^n} + \dots) = (\frac{r}{t})^{q_0} \frac{1}{1 - \frac{r}{t}} .$$

Thus the series $\sum_{q=0}^{+\infty} P_q(x)$ converges uniformly for $x \in \overline{B(0,r)}$, for every r satisfying (I.2.2).

Hence $R \geqslant \dfrac{1}{\limsup \|P_q\|^{1/q}}$, and that completes the proof of the lemma. Q.E.D.

Let c_{00} be the vector space of all sequences $x = (x_0, x_1, \dots, x_q, \dots)$ of complex numbers having only a finite number of non vanishing terms, with norm

$$\|x\| = \max_{n \in \mathbb{N}} |x_n| .$$

Its completion is the space c_0 of complex sequences which are converging to 0.

Consider the power series $\sum_{q=0}^{\infty} P_q$ from $\mathbb{C}$ to c_{00}, where $P_q(\zeta) = (0, \dots, \underset{q}{\zeta^q}, 0, \dots)$.

The series $\sum_{q=0}^{\infty} P_q$ has radius of convergence 1 in c_0, but in c_{00} the series $\sum_{q=0}^{\infty} P_q(\zeta)$ converges only for $\zeta=0$.

<u>Remark</u>. Let $\sum_{q=0}^{+\infty} P_q$ be a convergent power series from E to F, and let H be a linear subspace of E. Then $\sum_{q=0}^{+\infty} P_{q|H}$ is a convergent power series from H to F, and its radius of convergence is larger or equal to the radius of convergence of $\sum_{q=0}^{+\infty} P_q$.

Repeating word by word the proof of the so called Abel's lemma one obtains

<u>Lemma I.2.2</u>. *Let $E = \mathbb{C}$, let F be a Banach space and let $\sum_{q=0}^{+\infty} \zeta^q a_q$ be a formal power series from $\mathbb{C}$ to F. If $\sup\{|\zeta_0|^q \|a_q\| : q=0,1,2,\dots\} < \infty$ at some $\zeta_0 \in \mathbb{C}^\star$, then the power series $\sum_{q=0}^{+\infty} \zeta^q a_q$ is normally (hence uniformly) convergent on any compact subset of the open disc with center 0 and radius $|\zeta_0|$ in $\mathbb{C}$.*

From now on F will always be assumed to be complete.

By Lemma I.2.1, the radius of convergence R of the power series $\sum_{q=0}^{+\infty} P_q$ is also the radius of convergence of the scalar valued power series $\sum_{q=0}^{+\infty} \|P_q\| \zeta^q$.

Thus Abel's lemma yields the following definition of the radius of convergence R:

$$R = \sup \{r \in \mathbb{R}_+ : \sum_{q=0}^{+\infty} \|P_q\| r^q < \infty\}.$$

By (I.1.6)

$$\|f(x)\| \leqslant g(x) \qquad \text{for all } \quad x \in B(0,R),$$

where

$g: B(0,R) \to \mathbb{R}_+$ is the function defined by

$$g(x) = \sum_{q=0}^{+\infty} \|P(x)\| .$$

For any $\varepsilon > 0$ and for any $0 \leqslant r < R$ there is some index m_0 such that

$$\sum_{q=m+1}^{+\infty} \|P_q\| r^q < \varepsilon \qquad \text{whenever} \quad m \geqslant m_0 .$$

Thus, by (I.1.6),

$$\sum_{q=m+1}^{+\infty} \|P_q(x)\| \leqslant \sum_{q=m+1}^{+\infty} \|P_q\| \|x\|^q \leqslant \sum_{q=m+1}^{+\infty} \|P_q\| r^q < \varepsilon$$

for all $m \geqslant m_0$ and all $x \in \overline{B(0,r)}$. That proves

<u>Lemma I.2.3</u>. *For any* r *such that* $0 \leqslant r < R$, *the power series* $\sum_{q=0}^{+\infty} P_q$ *is normally convergent on* $\overline{B(0,r)}$.

The following corollary is a trivial consequence of Lemma I.2.1.

<u>Corollary I.2.4</u>. *The power series* $\sum_{q=0}^{+\infty} P_q$ *is convergent if,*

and only if, the set $\{\|P_q\|^{1/q} : q \in \mathbb{N}^\star\} \subset \mathbb{R}_+$ *is bounded.*

Lemma I.2.5. *Let* $A_q \in L_s^q(E,F)$ *be such that* $\hat{A}_q = P_q$. *The series* $\sum_{q=0}^{+\infty} P_q$ *is convergent if, and only if, the set* $\{\|A_q\|^{1/q} : q \in \mathbb{N}^\star\} \subset \mathbb{R}_+$ *is bounded.*

Proof. By Corollary I.2.4 we need only show that $\{\|P_q\|^{1/q}\}$ is bounded if, and only if, $\{\|A_q\|^{1/q}\}$ is bounded.

Inequalities (I.1.7) yield

$$\|P_q\|^{1/q} \leqslant \|A_q\|^{1/q} \leqslant \frac{q}{(q!)^{1/q}}\|P_q\|^{1/q} . \tag{I.2.3}$$

In view of the Stirling formula, we have

$$\lim_{q\to\infty} \frac{q!}{q^q e^{-q}\sqrt{2\pi q}} = 1 ,$$

i.e.

$$\lim_{q\to\infty} \frac{(q!)^{1/q}}{q} = \frac{1}{e} . \tag{I.2.4}$$

This formula, coupled with (I.2.3) and with the fact that the sequence $\dfrac{q}{(q!)^{1/q}}$ is increasing, completes the proof of the lemma. Q.E.D.

By (I.2.3) and (I.2.4) the number $\tilde{R}$ defined by

$$\frac{1}{\tilde{R}} = \limsup \|A_q\|^{1/q}$$

satisfies the inequalities

$$\frac{1}{R} \leqslant \frac{1}{\tilde{R}} \leqslant \frac{e}{R} ,$$

i.e.

$$R/e \leqslant \tilde{R} \leqslant R. \tag{I.2.5}$$

The number $\tilde{R}$ is called the *radius of restricted conver-*

gence of the power series $\sum_{q=0}^{+\infty} P_q$.

The following example will show that inequalities (I.2.5) cannot be improved in general.

<u>Example 1</u>. As in Example 2 of § 1, let $E = \ell_1(\mathbb{N}^\star), F = \mathbb{C}$, and let $P_q \in P^q(E, \mathbb{C})$ be the polynomial defined by $P_q(x) = \zeta^1 \dots \zeta^q$ $(x = (\zeta^1, \zeta^2, \dots) \in \ell_1(\mathbb{N}^\star))$.

By (I.1.9) and (I.2.4) the radius of convergence R of the power series $\sum_{q=0}^{+\infty} q!P_q$ is given by

$$\frac{1}{R} = \limsup\,(q!^{1/q}\|P_q\|^{1/q}) = \lim \frac{q!^{1/q}}{q} = \frac{1}{e},$$

i.e.

$$R = e.$$

If $A_q \in L_s^q(E, \mathbb{C})$ is defined by

$$A_q(x_1, \dots, x_q) = \frac{1}{q!} \Sigma\, \zeta_{j_1}^1 \dots \zeta_{j_q}^q ,$$

where $x_j = (\zeta_j^1, \zeta_j^2, \dots) \in \ell_1(\mathbb{N}^\star)$, and where the summation is extended over all permutations $(j_1, \dots, j_q)$ of $(1, \dots, q)$, then

$$\hat{A}_q = P_q ,$$

and, as in Example 2 of § 1,

$$\|\hat{A}_q\| = \frac{1}{q!} .$$

Thus the radius of strict convergence of the power series $\sum_{q=0}^{+\infty} q!\, P_q$ is $\tilde{R} = 1$.

<u>Example 2</u>. Let $E = F = \ell_2(\mathbb{N})$. Let $P_q \in P^q(E, E)$ be the polynomial defined by

$$P_q(x) = (0,0,\dots,0,\zeta_q^q,0,\dots) \qquad (q=0,1,\dots)$$

for $x = (\zeta_0,\zeta_1,\dots,\zeta_q,\dots) \in \ell_2(\mathbb{N})$.

Consider the power series

$$\sum_{q=0}^{\infty} P_p \ ;$$

its radius of convergence is given by

$$\frac{1}{R} = \limsup (\|P_q\|^{1/q}) = 1.$$

The same conclusion holds if we set $E = \ell_1(\mathbb{N})$ or $\ell_2(\mathbb{N})$ or $\ell_\infty(\mathbb{N})$ and $F = \ell_\infty(\mathbb{N})$, and define P_q in the same way.

We will now establish the principle of analytic continuation. We begin by proving

<u>Lemma I.2.6</u>. *Let* $a_q \in F$ $(q=0,1,\dots)$, *and assume that* $\sum_{q=0}^{+\infty} \zeta^q a_q = 0$ *for all* $\zeta \in \mathbb{C}$ *such that* $|\zeta| < r$ *for some* $r > 0$. *Then* $a_q = 0$ *for* $q=0,1,\dots$.

<u>Proof</u>. For $\zeta = 0$, we have $a_0 = 0$. Assuming $a_0 = a_1 = \dots = a_{q_0-1} = 0$ for some $q_0 \geqslant 1$, we shall show that $a_{q_0} = 0$.

Let $0 < s < r$. Since the series $\sum_{q=0}^{+\infty} s^q a_q$ converges, then $\lim_{q\to\infty} s^q a_q = 0$.

Hence, there is a finite constant $K > 0$ such that

$$s^q \|a_q\| \leqslant K \qquad \text{for } q=0,1,\dots .$$

Being

$$\sum_{q=q_0}^{+\infty} \zeta^q a_q = 0 \qquad \text{for } |\zeta| < r,$$

then

$$a_{q_0} = - \sum_{q=q_0+1}^{+\infty} \zeta^{q-q_0} a_q \qquad \text{for } 0 < |\zeta| < r.$$

Hence, for $0<|\zeta|<s$:

$$\|a_{q_0}\| \leqslant \sum_{q=q_0+1}^{+\infty} |\zeta|^{q-q_0}\|a_q\| \leqslant K \sum_{q=q_0+1}^{+\infty} \frac{|\zeta|^{q-q_0}}{s^q} =$$

$$= \frac{K}{s^{q_0+1}} |\zeta| (1+\frac{|\zeta|}{s}+(\frac{|\zeta|}{s})^2+\dots) = \frac{K}{s^{q_0}} \cdot \frac{|\zeta|}{s-|\zeta|} .$$

Letting $|\zeta|\to 0$ we obtain $a_{q_0} = 0$. Q.E.D.

<u>Proposition I.2.7</u>. *If there is some* $r>0$ *such that the power series* $\sum_{q=0}^{+\infty} P_q(x)$ *vanishes for every* $x \in B(0,r)$, *then* $P_q = 0$ *for* $q=0,1,2,\dots$.

<u>Proof</u>. Let $y \in E$, $y \neq 0$, and let $x = \zeta y$ for $\zeta \in \mathbb{C}$. Then $\sum_{q=0}^{+\infty} P_q(x) = \sum_{q=0}^{+\infty} \zeta^q P_q(y)$.

By Lemma I.2.6, $P_q(y) = 0$ for $q=0,1,\dots$. Q.E.D.

Consider the power series from E to F $\sum_{q=0}^{+\infty} P_q$, where $P_q = \hat{A}_q$, $\hat{A}_q \in L_s^q(E,F)$, $q=0,1,\dots$.

For $q\geqslant 1$, consider the symmetric $(q-1)$-linear map $A'_q \in L_s^{q-1}(E,L(E,F))$ defined by

$$A'_q(x_1,\dots,x_{q-1})\cdot v = q\, A_q(x_1,\dots,x_{q-1}, v) \qquad (x,v \in E),$$

and set $P'_q = \hat{A}'_q$.

<u>Remark 3</u>. A'_q and P'_q are the Fréchet-derivatives of A_q and P_q respectively (for the notion of Fréchet-derivative in normed vector space, see for example [J. Dieudonné,1]).

<u>Lemma I.2.8</u>. *For all* $q\geqslant 1$,

$$\|A'_q\| = q\|A_q\| ,$$

$$q\|P_q\| \leqslant \|P'_q\| \leqslant \frac{q^{q+1}}{q!}\|P_q\| .$$

Proof. The equality and the first inequality are obvious. The second inequality follows from the first and from (I.1.7).

Q.E.D.

Consider now the series $\sum_{q=1}^{\infty} P'_q$, from E to $L(E,F)$, which is called the *derivative* of the series $\sum_{q=0}^{\infty} P_q$.

Proposition I.2.9. *The power series* $\sum_{q=0}^{+\infty} P_q$ *and* $\sum_{q=1}^{\infty} P'_q$ *have the same radius* $\tilde{R}$ *of restricted convergence. In particular, if* $\sum_{q=0}^{\infty} P_q$ *is convergent, also* $\sum_{q=1}^{\infty} P'_q$ *is convergent.*

Moreover, setting $f(x) = \sum_{q=0}^{\infty} P_q(x)$ *and* $g(x) = \sum_{q=1}^{\infty} P'_q(x)$ *for all* $x \in B(0,\tilde{R})$, *we have*

$$df(x) = g(x), \qquad (x \in B(0,\tilde{R}))$$

where $df(x)$ *is the Fréchet-derivative of* f *at* x.

Proof. The first part of the proposition is a direct consequence of the equality in the preceding lemma.

The second part follows from the convergence to $f(x)$ of the first series and from the uniform convergence of the second to $g(x)$, on every ball $\overline{B(0,r)}$ with $0 \leqslant r < R$ (cf. e.g. [J. Dieudonné,1]). Q.E.D.

Corollary I.2.10. *Let* $\sum_{q=0}^{\infty} P_q$ *be a convergent power series from* E *to* F, *with radius of restricted convergence* $\tilde{R}$. *Let* $f(x) = \sum_{q=0}^{\infty} P_q(x)$ *for* $x \in B(0,\tilde{R})$.

Then, the function $f: B(0,\tilde{R}) \to F$ *is infinitely differentiable and*

$$\hat{d}^q f(0) = q!\, P_q \qquad (q=0,1,\dots).$$

Proof. The first statement is an obvious consequence of the

Proposition I.2.9. Hence, it suffices to prove the equality for $q=1$.

Since $\|P_q(x)\| \leqslant \|P_q\| \|x\|^q$, $q=0,1,$,

then

$$\|f(x) - P_0 - P_1(x)\| = \|\sum_{q=2}^{\infty} P_q(x)\| \leqslant \sum_{q=2}^{\infty} \|P_q(x)\| \leqslant$$

$$\leqslant \sum_{q=2}^{\infty} \|P_q\| \|x\|^q = \|x\|^2 \sum_{q=0}^{\infty} \|P_{q+2}\| \|x\|^q.$$

Q.E.D.

Remark. All results in this Chapter remain true, with the same definitions, for real normed and real Banach spaces.

Notes. Chapter I follows essentially §§ 1-4 of [L.Nachbin[1]] or the first pages of Douady's thesis [A. Douady, 1] . Theorem I.1.9 is in [S. Mazur-W. Orlicz, 1; p.179] , and for linear maps in [S. Banach, 1] .

Example 5 is due to F. Ricci.

CHAPTER II

HOLOMORPHIC FUNCTIONS

§ 1. Holomorphic functions.

Throughout this section E and F will be complex normed spaces.

Let U be an open subset of E. A mapping $f: U \to F$ is called a *holomorphic function* on U with values in F, or an *F-valued holomorphic function* if, for every $u \in U$ there exist an open neighborhood V of u in U and a convergent power series $\sum_{q=0}^{+\infty} P_q$ with radius of convergence $R > 0$ such that, on $V \cap B(u,R)$, f is expressed by

$$(II.1.1) \qquad f(x) = \sum_{q=0}^{+\infty} P_q(x-u) \qquad (x \in V \cap B(u,R)).$$

Remarks. 1. By the definition in I.2, the convergence is uniform on $V \cap B(u,r)$ for every $0 < r < R$. Thus *a holomorphic function is continuous.*

2. By Proposition I.2.7 the power series $\sum_{q=0}^{+\infty} P_q$ defined by f and u is unique.

It will be called the *power series expansion of* f *at* u or *around* u; P_1 will be called the *linear part of* f *at* u.

3. For every open set $W \in U$, the restriction $f_{|W}$ of f to W is a holomorphic function with values in F. In other words, holomorphy is a local property.

Let $Hol(U,F)$ be the set of all F-valued holomorphic functions on U. It is a complex vector space under pointwise composition. By Remark 3, for every open subset W of U, the identity map $W \hookrightarrow U$ defines a linear map $Hol(U,F) \to Hol(W,F)$. By Proposition I.2.7, this linear map is injective if U is connected.

Examples. 1. Let P_q be a q-homogeneous polynomial from E to F. For any $u \in E$, we have, by the Newton identity,

$$P_q(u+y) = \sum_{p=0}^{q} R_p(y) \qquad (y \in E),$$

where R_p is a p-homogeneous polynomial from E to F. Each R_p is continuous if, and only if, P is continuous. This shows that *a polynomial from E to F is holomorphic if, and only if, it is continuous.*

Hence

$$P(E,F) \subset Hol(E,F).$$

2. Let A be a complex Banach algebra with an identity e and let $A^\star$ be the set of all invertible elements of A. Let $f: A^\star \to A$ be the function defined by $f(x) = x^{-1}$.

For any $u \in A^\star$ and all $x \in A$

$$u + x = (e + xu^{-1})u.$$

If $\|xu^{-1}\| < 1$ - hence, *a fortiori*, if $\|x\| < \|u^{-1}\|^{-1}$ - then

$$\left\| \sum_{q=0}^{+\infty} (-1)^q (xu^{-1}) \right\| \leqslant \sum_{q=0}^{+\infty} \|xu^{-1}\| = \frac{1}{1-\|xu^{-1}\|}.$$

Thus, for any r, with $0 \leqslant r < \frac{1}{\|u^{-1}\|}$, the series $\sum_{q=0}^{+\infty} (-1)^q u^{-1}(xu^{-1})^q$ converges uniformly for $x \in \overline{B(0,r)}$, and its sum is $(u+x)^{-1}$.

That proves the (well known) fact that $A^\star$ is open. Let P_q be the q-homogeneous polynomial from A to A defined by

(II.1.2) $$P_q(x) = (-1)^q u^{-1}(xu^{-1}) \qquad (q=0,1,\dots)$$

and let A_q be the q-linear map of A^q into A defined by

(II.1.3) $$A_q(x_1,\dots,x_q) = (-1)^q u^{-1}x_1u^{-1}x_2\dots u^{-1}x_qu^{-1}.$$

Then

$$\|A_q(x_1,\dots,x_q)\| \leqslant \|u^{-1}\|^{q+1}\|x_1\|\dots\|x_q\| ,$$

and therefore $A_q \in L^q(A,A)$, and

(II.1.4) $$\|A_q\| \leqslant \|u^{-1}\|^{q+1}.$$

Since $\hat{A}_q = P_q$, then $P_q \in P^q(A,A)$. For $\|x\| < \|u^{-1}\|^{-1}$

$$f(u+x) = (u+x)^{-1} \qquad \sum_{q=0}^{+\infty} P_q(x).$$

Thus f is holomorphic on $A^\star$, $f \in Hol(A^\star,A)$.

Computing $\|P_q\|$ we have

II.1.5) $$\|P_q\| = \sup\{\|P_q(x)\|: \|x\| \leqslant 1\} \geqslant \|P_q(e)\| = \|(u^{-1})^{q+1}\|.$$

Being

$$\|(u^{-1})^{q+1}\| = \|(u^{q+1})^{-1}\| \geqslant \|u^{q+1}\|^{-1} \geqslant \|u\|^{-(q+1)},$$

then

$$\|P_q\| \geqslant \|u\|^{-(q+1)},$$

and the radius R of convergence of $\sum_{q=0}^{+\infty} P_q$ is given by

$$\frac{1}{R} = \lim\sup \|P_q\|^{1/q} \geqslant \|u\|^{-1},$$

i.e.

$$R \leqslant \|u\|.$$

As for the radius of restricted convergence $\tilde{R}$, (II.1.4) yields

$$\frac{1}{\tilde{R}} \leqslant \lim\sup\|(A_q)_s\| \leqslant \lim\sup\|A_q\|^{1/q} \leqslant \|u^{-1}\|,$$

where $(A_q)_s$ is the q-linear symmetric operator associated to A_q.

Then

$$\tilde{R} \geqslant \|u^{-1}\|^{-1},$$

and so

$$\|u^{-1}\|^{-1} \leqslant \tilde{R} \leqslant R \leqslant \|u\|.$$

3. If the complex Banach algebra A is commutative, then the polynomial P_q expressed by (II.1.2) and (II.1.3) and the map A_q are given by

$$P_q(x) = (-1)^q u^{-(q+1)} x^q, \; A_q(x_1, \dots, x_q) = (-1)^q u^{-(q+1)} x_1 \dots x_q.$$

Hence $A_q \in L_s^q(A, A)$, and

$$\|A_q\| \leqslant \|u^{-(q+1)}\|.$$

Since $\|P_q\| \leqslant \|A_q\|$, the above inequality and (II.1.5) yield

$$\|A_q\| = \|P_q\| = \|u^{-(q+1)}\|.$$

Hence

$$\frac{1}{R} = \frac{1}{\tilde{R}} = \lim\sup\|u^{-(q+1)}\|^{1/q} = \lim(\|(u^{-1})^{q+1}\|^{1/(q+1)})^{(q+1)/q} = \\ = \rho(u^{-1}),$$

where ρ is the spectral radius.

4. Going back to Example 2, let D be a domain in $\mathbb{C}$, whose boundary Γ consists of a finite number of closed disjoint rectifiable Jordan curves. By the upper semi-continuity of the spectrum (cf. e.g. [Hille and Phillips, 1, p.167]), the set

$$U = \{x \in A: \operatorname{Sp} x \subset D\}$$

is open in A. Let f be a scalar-valued holomorphic function on a neighborhood of the closure $\overline{D}$ of D. Let $x \in U$, and let $f(x)$ be the element of A defined by the Dunford integral

$$f(x) = \frac{1}{2\pi i} \int_\Gamma f(\zeta)(\zeta e-x)^{-1} d\zeta$$

(cf. e.g. [Hille and Phillips, 1, pp. 164-211]), where the path of integration is oriented counterclockwise. Then, by the spectral mapping theorem, $\operatorname{Sp} f(x) = f(\operatorname{Sp} x) \subset D$.

We will show that the function $x \mapsto f(x)$ is a holomorphic map of U into A. For any $u \in U$ and any $\zeta \in \Gamma$ $\zeta e-u$ is invertible.

Thus, by Example 2, for any $x \in A$ such that

$$\|x\| < \|(\zeta e-u)^{-1}\|^{-1},$$

the series

$$\text{(II.1.6)} \qquad \sum_{q=0}^{+\infty} (-1)^q (\zeta e-u)^{-1} (x(\zeta e-u)^{-1})^q$$

converges to $(\zeta e-(u+x))^{-1}$ uniformly for $\|x\| \leqslant r$, for any $0 \leqslant r < \|(\zeta e-u)^{-1}\|^{-1}$. Hence integration term by term on (II.1.6) yields

$$f(u+x) = \sum_{q=0}^{+\infty} P_q(x)$$

where $P_q \in \mathcal{P}^q(A,A)$ is defined by

$$P_q(x) = \frac{1}{2\pi i} \int_\Gamma f(\zeta)(\zeta e-u)^{-1} (x(\zeta e-u)^{-1})^q d\zeta.$$

Clearly $P_q = \hat{A}_q$, where A_q is the q-linear map of A into A defined on $x_1, \dots, x_q \in A$ by the integral

$$A_q(x_1,\dots,x_q) = \frac{1}{2\pi i} \int_\Gamma f(\zeta)(\zeta e-u)^{-1} x_1 (\zeta e-u)^{-1} \dots x_q (\zeta e-u)^{-1} d\zeta.$$

Let $k = \frac{1}{2\pi} \max_\Gamma |f| \int_\Gamma |d\zeta|$. Since

$$\|A_q(x_1, \dots, x_q)\| \leqslant k \|(\zeta e-u)^{-1}\|^{q+1} \|x_1\| \dots \|x_q\|,$$

for all $x_1, \dots, x_q \in A$, then $A_q \in \mathcal{L}^q(A,A)$, and

$$\|A_q\| \leq k\|(\zeta e-u)^{-1}\|^{q+1}$$

Thus the radius of restricted convergence $\tilde{R}$ of the power series $\sum_{q=0}^{+\infty} P_q$ is

$$\frac{1}{\tilde{R}} \leq \|(\zeta e-u)^{-1}\|.$$

By (I.2.5) the power series $\sum_{q=0}^{\infty} P_q$ has radius of convergence

$$R \geq \|(\zeta e-u)^{-1}\|^{-1} > 0,$$

and in conclusion $f \in \mathrm{Hol}(U,A)$.

§ 2. The inverse mapping theorem.

Let E, F and H be complex normed spaces. Let U and W be open sets in E and F, and let $f \in \mathrm{Hol}(U,F)$, $h \in \mathrm{Hol}(W,H)$ be such that the open set $U \cap f^{-1}(W)$ is not empty. We will prove that $h \circ f \in \mathrm{Hol}(U \cap f^{-1}(W),H)$. We begin by computing the power series expansion of $h \circ f$ around any point $u \in U \cap f^{-1}(W)$ in terms of the power series expansions of f around u and of h around $f(u)$.

There is no restriction in assuming $u = 0$, $f(u) = 0$, $h(0) = 0$. Let

$$\text{(II.2.1)} \qquad f(x) = \sum_{q=1}^{+\infty} \hat{A}_q(x), \qquad h(y) = \sum_{q=1}^{+\infty} \hat{B}_q(y),$$

- with $A_q \in \mathcal{L}_s^q(E,F)$, $B_q \in \mathcal{L}_s^q(F,H)$ - be the power series expansions of f and h around 0.

The proof of the following lemma will be left as an excercise:

<u>Lemma II.2.1</u>. *Let* T *be a continuous* p-*linear map of* F^p

into H, *and let* $P^1, \dots, P^p$ *be continuous homogeneous polynomials from* E *to* F *with degrees* $q_1, \dots, q_p$. *The function* $x \mapsto T(P^1(x), \dots, P^p(x))$ *is a continuous homogeneous polynomial from* E *to* H *with degree* $q_1 + \dots + q_p$.

Let $C_q \in L_s^q(E,H)$ be the operator defined by the q-homogeneous polynomial $\hat{C}_q \in P^q(E,H)$ expressed by

$$\text{(II.2.2)} \qquad \hat{C}_p(x) = \sum_{p=1}^{q} \sum_{\substack{r_1+\dots+r_p=q \\ r_1 \geqslant 1, \dots, r_p \geqslant 1}} B_p(\hat{A}_{r_1}(x), \dots, \hat{A}_{r_p}(x))$$

for $q = 1,2,\dots$. We will prove that $\sum_{q=1}^{+\infty} \hat{C}_q$ is a convergent power series and that

$$\text{(II.2.3)} \qquad h(f(x)) = \sum_{q=1}^{+\infty} \hat{C}_q(x)$$

in a neighborhood of 0. Due to the normal convergence of the power series (II.2.1) near 0, to establish this result we need only show that the radius of convergence of the power series $\sum_{q=1}^{+\infty} \hat{C}_q$ is positive. Let $R(f)$, $R(h)$ and $\tilde{R}(f)$, $\tilde{R}(h)$ be the radii of convergence and the radii of restricted convergence of the power series (II.2.1).

For any $t \geqslant 0$ we set

$$\rho(f)_t = \sum_{q=1}^{+\infty} \|\hat{A}_q\| t^q,$$

$$\tilde{\rho}(f)_t = \sum_{q=1}^{+\infty} \|A_q\| t^q,$$

and we introduce similar notations for h. Note that

$$\rho(f)_t < \infty \quad \text{for} \quad 0 \leqslant t < R(f), \quad \rho(f)_t = +\infty \quad \text{for} \quad t > R(f),$$

$$\tilde{\rho}(f)_t < \infty \quad \text{for} \quad 0 \leqslant t < \tilde{R}(f), \quad \tilde{\rho}(f)_t = +\infty \quad \text{for} \quad t > \tilde{R}(f),$$

etc. Since, by (II.2.2),

$$\|\hat{C}_q\| \leqslant \sum_{p=1}^{q} \|B_p\| \sum_{\substack{r_1+\dots+r_p=q \\ r_1\geqslant 1,\dots,r_p\geqslant 1}} \|\hat{A}_{r_1}\| \dots \|\hat{A}_{r_p}\|,$$

then, for any $t \geqslant 0$,

$$\sum_{q=1}^{+\infty} \|C_q\| t^q \leqslant \sum_{q=1}^{+\infty} \sum_{p=1}^{q} \|B_p\| \sum_{\substack{r_1+\dots+r_p=q \\ r_1\geqslant 1,\dots,r_p\geqslant 1}} \|\hat{A}_{r_1}\| t^{r_1} \dots \|\hat{A}_{r_p}\| t^{r_p} = \tag{II.2.4}$$

$$= \|B_1\| (\|\hat{A}_1\| t + \|\hat{A}_2\| t^2 + \dots) + \|B_2\| (\|\hat{A}_1\| t + \|\hat{A}_2\| t^2 + \dots)^2 + \dots$$

$$= \|B_1\| \rho(f)_t + \|B_2\| (\rho(f)_t)^2 + \dots = \tilde{\rho}(h)_{\rho(f)_t}.$$

That proves that the power series $\Sigma \hat{C}_p$ has a positive radius of convergence and thereby (II.2.3) holds uniformly in a neighborhood of 0 in E. Hence $h \circ f$ is holomorphic in a neighborhood of 0, and (II.2.4) reads now

$$\rho(h \circ f)_t \leqslant \tilde{\rho}(h)_{\rho(f)_t}. \tag{II.2.5}$$

This inequality and the following well known lemma will be crucial in the proof of the inverse mapping theorem.

Lemma II.2.2. *Let* (f_ν) *be a sequence of (complex valued) equi-bounded holomorphic functions on* $\Delta(0,\rho)=\{Z \in \mathbb{C} : |z|<\rho\}$, *and such that, for* $n = 1,2,\dots,$ *the sequence* $(f_\nu^{(n)}(0))$ *is convergent. Then* (f_ν) *is uniformly convergent in* $\Delta(0,r)$ *for all* r *such that* $0 \leqslant r < \rho$.

Proof. Let $M > 0$ be such that $|f_\nu(z)| \leqslant M \;\; \forall\, z \in \Delta(0,\rho), \; \forall\, \nu$. By the Cauchy inequality for scalar valued functions of one variable (cf. e.g. [Ahlfors, 1]),

$$\frac{1}{n!} |f_\nu^{(n)}(0)| \leqslant \frac{M}{\rho^n}.$$

Let $0 < r < \rho$ and $|z| \leqslant r$. Then, for any positive integer N,

$$|f_\nu(z) - f_\mu(z)| \leqslant \sum_{n \leqslant N} \frac{1}{n!} |f_\nu^{(n)}(0) - f_\mu^{(n)}(0)| r^n +$$

$$+ \sum_{n > N} \frac{1}{n!} |f_\nu^{(n)}(0)| r^n + \sum_{n > N} \frac{1}{n!} |f_\mu^{(n)}(0)| r^n \leqslant$$

$$\leqslant \sum_{n \leqslant N} \frac{1}{n!} |f_\nu^{(n)}(0) - f_\mu^{(n)}(0)| r^n + 2M \sum_{n > N} \left(\frac{r}{\rho}\right)^n \leqslant$$

$$\leqslant \sum_{n \leqslant N} \frac{1}{n!} |f_\nu^{(n)}(0) - f_\mu^{(n)}(0)| r^n + 2M \left(\frac{r}{\rho}\right)^{N+1} \frac{1}{1 - \frac{r}{\rho}} .$$

Given $\varepsilon > 0$, there exists $N > 0$ such that

$$2M \left(\frac{r}{\rho}\right)^{N+1} \frac{1}{1 - \frac{r}{\rho}} < \varepsilon ,$$

and there exists ν_0 such that $\nu, \mu > \nu_0$ implies

$$\frac{1}{n!} |f_\nu^{(n)}(0) - f_\mu^{(n)}(0)| r^n < \varepsilon/_{N+1} \quad \text{for} \quad n \leqslant N.$$

Then, for $|z| \leqslant r$ and $\mu, \nu > \nu_0$

$$|f_\nu(z) - f_\mu(z)| \leqslant 2\varepsilon$$

Q.E.D.

Theorem II.2.3. *Let* U *be an open set in* E *and let* $f: U \to F$ *be a holomorphic function. If the linear part of the power series expansion* f *at some point* $u \in U$ *is a bi-continuous isomorphism of* E *onto* F, *then there exists an open neighborhood* V *of* u *in* U *such that* $f(V)$ *is open in* F, $f_{|V}$ *is a homeomorphism of* V *onto* $f(V)$ *and* $(f_{|V})^{-1}$ *is holomorphic on* $f(V)$.

Proof. 1. As before, we assume $u = 0$, $f(u) = 0$. Let

$$f(x) = \sum_{q=1}^{+\infty} P_q(x)$$

be the power series expansion of f in an open neighborhood W of 0. In view of the hypothesis, the linear part P_1 of f at 0 is a bi-continuous isomorphism of E onto F. Let $g: U \to E$ be the function defined by

(II.2.6) $$g(x) = x - P_1^{-1} \circ f(x).$$

Then g is holomorphic and its power series expansion in a neighborhood of 0 is

$$g(x) = -(P_1^{-1} \circ P_2(x) + P_1^{-1} \circ P_3(x) + \dots).$$

Since the linear part of g at 0 is zero, for any k, with $0 < k < 1$, there is some $t > 0$ such that

(II.2.7) $$\tilde{\rho}(g)_s \leqslant ks \quad \text{for all} \quad 0 \leqslant s \leqslant t.$$

2. Let h $(n = 0, 1, \dots)$ be a holomorphic map of an open neighborhood of $0 \in F$ into E, defined inductively by

(II.2.8) $$h_0(y) = 0, \quad h_1(y) = P_1^{-1}(y), \dots, h_{n+1}(y) = P_1^{-1}(y) + g \circ h_n(y).$$

Let $t' = \dfrac{(1-k)t}{\|P_1^{-1}\|}$. We will show that

(II.2.9) $$\rho(h_n)_{t'} \leqslant t \quad \text{for} \quad n = 0, 1, 2, \dots$$

If $n=1$, then

$$\rho(h_1)_{t'} = \|P_1^{-1}\| t' = (1-k)t < t.$$

If (II.2.9) holds for n, then by (II.2.8), (II.2.7) and (II.2.9),

$$\rho(h_{n+1})_{t'} \leqslant \|P_1^{-1}\| t' + \rho(g \circ h_n)_{t'} \leqslant (1-k)t + \tilde{\rho}(g)_{\rho(h_n)_{t'}} \leqslant$$

$$\leqslant (1-k)t + k\rho(h_n)_{t'} \leqslant (1-k)t + kt = t.$$

That proves (II.2.9). Hence h_n is holomorphic on the ball $B_F(0,t')$. Since

$$h_{n+1} - h_n = g \circ h_n - g \circ h_{n-1}$$

the fact that the linear part of g at 0 is zero, implies that $h_{n+1}-h_n$ has *order* $> n$ at 0, i.e. the terms of degree $\leqslant n$ in the power series expansion of $h_{n+1}-h_n$ in an open neighborhood of 0 in E all vanish.

Now, let $\sum_{q=0}^{+\infty} P_q^{(n)}$ be the power series expansion of h_n at 0, and let us consider the power series $\sum_{q=0}^{+\infty} \|P_q^{(n)}\| z^q$ $(z \in \mathbb{C})$.

Being $\rho(h_n)_{t'} \leqslant t \ \forall\, n$, the power series $\sum_{q=0}^{+\infty} \|P_q^{(n)}\| z^q$ defines a sequence of functions which are uniformly bounded on the disc $\Delta(0,t')$. By Lemma II.2.2, the power series $\sum_{q=0}^{+\infty} \|P_q^{(q)}\| z^q$ defines a holomorphic function on $\Delta(0,t')$, and therefore the sequence (h_n) defines a holomorphic function h on an open neighborhood of 0 on F, such that

$$\rho(h)_{t'} \leqslant t.$$

3. For any $y \in B_F(0,t')$, (II.2.8) and (II.2.6) yield

$$P_1 \circ h_{n+1}(y) = y + P_1 \circ g \circ h_n(y) =$$

$$= y + P_1 \circ (h_n(y) - P_1^{-1} \circ f \circ h_n(y)) =$$

$$= y + P_1 \circ h_n(y) - f \circ h_n(y),$$

i.e.

$$P_1 \circ (h_{n+1}(y) - h_n(y)) = y - f \circ h_n(y).$$

Since $h_{n+1}-h_n$ has order $> n$ at 0, then also $Id - f \circ h_n$ has order $> n$ at 0. Thus

$$f \circ h = Id \qquad \text{on } B_F(0,t'),$$

i.e. f has a right inverse which is holomorphic on an open neighborhood of 0 in E.

Applying this conclusion to h we construct a right inverse of h which is holomorphic on a neighborhood of 0 in F. Q.E.D.

Remark. If E is a Banach space it suffices to assume the linear part of f at 0 to be bi-jective and continuous.

§ 3. Taylor expansion.

Let U be an open set in a complex normed space E, let F be a complex Banach space, and let $f \in \mathrm{Hol}(U,F)$. If E_0 is an affine subspace of E, then the restriction $f_{|U\cap E_0}$ is holomorphic, $f_{|U\cap E_0} \in \mathrm{Hol}(U\cap E_0, F)$.

If λ is a continuous linear form on F, then $\lambda\circ f$ is a complex-valued holomorphic function U: $\lambda\circ f \in \mathrm{Hol}(U,\mathbb{C})$.

We shall begin by discussing briefly F-valued holomorphic functions of one complex variable. A direct application of the Hahn-Banach theorem yields immediately the Cauchy integral theorem and the Cauchy integral formula.

Proposition II.3.1. *Let* U *be open in* $\mathbb{C}$. *If* $f \in \mathrm{Hol}(U,F)$ *and if* ℓ *is a closed rectifiable curve* ℓ *in* U, *which is homotopic to a point in* U, *then*

$$\int_{\ell} f(\zeta)\,d\zeta = 0.$$

Proposition II.3.2. *Let* $f \in \mathrm{Hol}(U,F)$ (U *open in* $\mathbb{C}$) *and let* ℓ *be a closed rectifiable curve in* U. *Then for every* $z \in U$ *which does not lie on* ℓ

$$I(z,\ell)f(z) = \frac{1}{2\pi i}\int_\ell \frac{1}{\zeta - z} f(\zeta)d\zeta ,$$

where $I(z,\ell)$ *is the index of* ℓ *with respect to* z.

For the definition of the index cf. e.g. [Ahlfors,1,pp.114-119].

Going back to the general case, let U be an open set in the complex normed space E and let $f \in Hol(U,F)$.

For any $u \in U$ and any $y \in E$, the set

$$U_{u,y} = \{\zeta \in \mathbb{C}: u+\zeta y \in U\}$$

is an open subset of $\mathbb{C}$, and the function $f_{u,y}: U_{u,y} \to F$ defined by

$$f_{u,y}(\zeta) = f(u+\zeta y)$$

is holomorphic on $U_{u,y}$. Thus Propositions II.3.1 and II.3.2 apply. As an example, we have

Lemma II.3.3. *Let* $u \in U$ *and let* $y \in E$ *be such that* $u+\zeta y \in U$ *for all* $|\zeta| < r$ *for some* $r > 1$. *Then*

$$f(u+y) = \frac{1}{2\pi i}\int_{|\zeta|=r} \frac{1}{\zeta-1} f(u+\zeta y)d\zeta .$$

Here, as always in the following, the integration-path is oriented counter-clockwise.

Now, let $\sum_{q=0}^{+\infty} P_q$ be the power series expansion of f at u, with radius of convergence $R > 0$, and let V be an open neighborhood of u in U such that (II.1.1) holds for all $x \in V \cap B(u,R))$.

Let $A_q \in L_s^q(E,F)$ be such that $\hat{A}_q = P_q$. In the following we shall set

$$d^q f(u) = q!A_q , \qquad \hat{d}^q f(u) = q!P_q ,$$

so that (II.1.1) takes the form of the *Taylor expansion at* u

$$f(x) = \sum_{q=0}^{+\infty} \frac{1}{q!} \hat{d}^q f(u)(x-u) \qquad (x \in V \cap B(u,r)). \tag{II.3.1}$$

Choosing r so small that $0 < r < R$ and $\overline{B(u,r)} \subset V$, in view of the uniform convergence of $\sum_{q=0}^{+\infty} P_q$ on $\overline{B(u,r)}$, the value of $P_q = \frac{1}{q!} \hat{d}^q f(u)$ on any $y \in E$, $\|y\| \leqslant r$, is given by

$$\frac{1}{q!} \hat{d}^q f(u)(y) = \frac{1}{2\pi i} \int_{|\zeta|=1} \frac{1}{\zeta^{q+1}} f(u+\zeta y)\,d\zeta = \tag{II.3.2}$$

$$= \frac{1}{2\pi} \int_0^{2\pi} e^{-iq\theta} f(u+e^{i\theta} y)\,d\theta,$$

i.e. for any $y \in E\setminus\{0\}$ we have

$$\frac{1}{q!} \hat{d}^q f(u)(y) = \frac{\|y\|^q}{2\pi r^q} \int_{|\zeta|=1} \frac{1}{\zeta^{q+1}} f\left(u + \frac{r\zeta}{\|y\|} y\right) d\zeta = \tag{II.3.3}$$

$$= \frac{\|y\|^q}{2\pi r^q} \int_0^{2\pi} e^{-iq\theta} f\left(u + \frac{re^{i\theta}}{\|y\|} y\right) d\theta .$$

For $q=0$ and $\|y\| \leqslant r$, (II.3.2) yields

$$f(u) = \frac{1}{2\pi} \int_0^{2\pi} f(u + e^{i\theta} y)\,d\theta ,$$

whence

$$\|f(u)\| \leqslant \frac{1}{2\pi} \int_0^{2\pi} \|f(u + e^{i\theta} y)\|\,d\theta .$$

Since f is continuous on U, this inequality implies

<u>Theorem II.3.4</u>. *(Maximum principle). If* $f \in \mathrm{Hol}(U,F)$ *the function* $x \mapsto \|f(x)\|$ *can have no maximum at* $x_0 \in U$ *unless* $\|f(\cdot)\|$ *is constant on a neighborhood of* x_0.

Let s be such that $0 < r < s < R$, and $\overline{B(u,s)} \subset U$. The function $\zeta \mapsto \frac{1}{\zeta^{q+1}} f(u + \frac{\zeta}{\|y\|} y)$ $(y \in B(u,s))$ is holomorphic on an open neighborhood of the closed annulus $\{\zeta \in \mathbb{C}: r \leqslant \zeta \leqslant s\}$. Thus, by Proposition II.3.1

$$\int_{|\zeta|=r} \frac{1}{\zeta^{q+1}} f(u + \zeta y)d\zeta = \int_{|\zeta|=s} \frac{1}{\zeta^{q+1}} f(u + \zeta y)d\zeta,$$

and that proves

Lemma II.3.5. *The coefficient* $\hat{d}^q f(u)$ *of the Taylor expansion* (II.3.1) *of* f *at* u *is given by* (II.3.3), *where* $y \in E$ *and* r *is any positive number such that* $\overline{B(u,r)} \subset U$.

As a consequence of Theorem II.3.4 and of (I.1.7) we have

Proposition II.3.6. *(Cauchy inequalities). For* $u \in U$ *let* $d(u)$ *be the distance of* u *from* $E\setminus U$. *Then for any* r *such that* $0 < r \leqslant d(u)$

$$\| \hat{d}^q f(u) \| \leqslant \frac{q!}{r^q} \sup\{\| f(x) \| : x \in B(u,r)\}$$

$$\| d^q f(u) \| \leqslant \frac{q^q}{r^q} \sup\{\| f(x) \| : x \in B(u,r)\}.$$

Lemma II.3.7. *Given* $u \in U$, *let* $y \in E$ *be such that* $u + \zeta y \in U$ *for all* $|\zeta| \leqslant r$ *and for some* $r > 1$. *Then, for* $n = 0,1,2,\dots,$

$$\| f(u+y) - \sum_{q=0}^{n} \frac{1}{q!} \hat{d}^q f(u)(y) \| \leqslant \frac{1}{r^n(r-1)} \sup\{\| f(u+\zeta y) \| : |\zeta|=r\}.$$

Proof. Lemmas II.3.3 and II.3.5 yield

$$f(u+y) - \sum_{q=0}^{n} \frac{1}{q!} \hat{d}^q f(u)(y) = \frac{1}{2\pi i} \int_{|\zeta|=r} \left(\frac{1}{\zeta-1} - \sum_{q=0}^{n} \frac{1}{\zeta^{q+1}} \right) f(u+\zeta y)d\zeta =$$

$$= \frac{1}{2\pi i} \int_{|\zeta|=r} \frac{1}{\zeta^{n+1}(\zeta-1)} f(u+\zeta y)d\zeta .$$

Since, for $|\zeta| = r$, $|\zeta-1| \geqslant r-1$, then

$$\| f(u+y) - \sum_{q=0} \frac{1}{q!} \hat{d}^q f(u)(y) \| \leqslant \frac{1}{2\pi} \int_{|\zeta|=r} \frac{1}{|\zeta|^{n+1}|\zeta-1|} \| f(u+\zeta y) \| |d\zeta| \leqslant$$

$$\leqslant \frac{1}{r^n(r-1)} \sup\{\| f(u+\zeta y) \| : |\zeta| = r\}.$$

Q.E.D.

We discuss now a Liouville-type theorem characterizing continuous polynomials. First of all, by Example 1, $\mathcal{P}^q(E,F) \subset$ $\subset \mathrm{Hol}(E,F)$. If $P_q \in \mathcal{P}^q(E,F)$, then

$$\hat{d}^n P_q(0)(x) = \frac{n!}{2\pi} \int_0^{2\pi} e^{-in\theta} P_q(e^{i\theta}x)\,d\theta =$$

$$= \frac{n!}{2\pi} \left(\int_0^{2\pi} e^{i(q-n)\theta} d\theta \right) P_q(x) =$$

$$= \begin{cases} 0 & \text{if } n \neq q, \\ q!\,P_q(x) & \text{if } n=q. \end{cases}$$

Viceversa, let $f \in \mathrm{Hol}(U,F)$, with U connected, and suppose there is some u and some integer $q_0 \geq 0$ such that $d^q f(u) = 0$ for all $q > q_0$. Then f is the restriction to U of a continuous polynomial of degree $\leq q_0$. In conclusion, we have

<u>Lemma II.3.8</u>. *Let* $f \in \mathrm{Hol}(U,F)$, *where* U *is a domain in* E. *Then* f *is the restriction to* U *of a continuous polynomial of degree* $\leq q_0$ *if, and only if, there is some point* $u \in D$ *such that* $d^q f(u) = 0$ *for all* $q > q_0$.

A continuous polynomial $P \neq 0$ *is a* q-*homogeneous continuous polynomial if, and only if,* $d^n P(0) = 0$ *for all* $n \neq q$, *while* $d^q P(0) \neq 0$.

The following Liouville-type theorem follows from Cauchy inequalities and from Lemma II.3.7 as in the classical case.

<u>Theorem II.3.9</u>. *Let* $f \in \mathrm{Hol}(E,F)$. *If there is an integer* $q_0 \geq 0$ *and a constant* $k > 0$ *such that* $\|f(x)\| \leq k\|x\|^{q_0}$ *for all* $x \in E$, *then* f *is a continuous polynomial of degree* $\leq q_0$. *In particular, if* f *is bounded,* f *is constant.*

The following theorem, which will be proved in the next

section, gives a characterization of Banach valued holomorphic functions in terms of complex-valued functions of one complex variable. Denoting by F' the topological dual of F, recall that a *determining manifold* for F is a closed subspace F_0' of F' such that, for every $y \in F$,

$$\|y\| = \sup\{|\lambda(y)| : \lambda \in F_0', \ \|\lambda\| \leqslant 1\}.$$

Theorem II.3.10. *Let* F_0' *be a determining manifold for* F. *A function* $f: U \to F$ *is holomorphic if, and only if,* f *is locally bounded, and for every* $\lambda \in F_0'$ *and every complex affine line* $E_0 \subset E$, *the complex-valued function* $\lambda \circ f$ *is holomorphic on* $E_0 \cap U$.

A first consequence of this theorem is the following proposition, which generalizes Example 1.

Proposition II.3.11. *Let* $\sum_{q=0}^{+\infty} P_q$ *be a power series from* E *to* F *with radius of convergence* $R > 0$. *The sum of the series is holomorphic on* $B(0,R)$.

As a consequence of Theorem II.3.10 we prove now

Proposition II.3.12. *For any* $f \in \mathrm{Hol}(U,F)$ *the functions* $\hat{d}^q f: u \mapsto \hat{d}^q f(u)$ *and* $d^q f: u \mapsto d^q f(u)$ *are holomorphic:*

$$\hat{d}^q f \in \mathrm{Hol}(U, P^q(E,F)), \quad d^q f \in \mathrm{Hol}(U, L_s^q(E,F)).$$

Proof. By Proposition I.1.3 we need consider only $\hat{d}^q f$.

Let $\lambda \in F'$ be any continuous linear form on F and let $y \in E$. The map $(\lambda,y): P \to \lambda \circ P(y)$ is a continuous linear form on $P^q(E,F)$. Thus there is an imbedding of $F' \times E$ as a closed subspace of the dual space $(P^q(E,F))'$. Since the norm $\|(\lambda,y)\|$ of the linear form (λ,y) is

$$\|(\lambda,y)\| = \sup\{|\lambda(P(y))|: \|P\| \leqslant 1\} \leqslant \|\lambda\|\|y\|^q,$$

the image of the product of the unit balls in F' and E is contained in the unit ball of $(P^q(E,F))'$. For any $P \in P^q(E,F)$,

$$\sup\{|(\lambda,y)(P)|:\|\lambda\| \leqslant 1,\|y\| \leqslant 1\} = \sup\{|\lambda(P(y))|:\|\lambda\| \leqslant 1,\|y\| \leqslant 1\} =$$

$$= \sup\{\|P(y)\|:\|y\| \leqslant 1\} = \|P\|.$$

Thus $F' \times E$ is a determining manifold for $P^q(E,F)$.

By the Cauchy inequalities (Proposition II.3.6), $\hat{d}^q f$ and $d^q f$ are locally bounded; thus, by Theorem II.3.10, we need only show that for any $\lambda \in F'$, and for any choice of u,y,x in E, the function $\zeta \mapsto \lambda(\hat{d}^q f(u+\zeta x)(y))$ is holomorphic in the open set $U_{u,x} = \{\tau \in \mathbb{C}: u+\tau x \in U\} \subset \mathbb{C}$.

Let Γ be a simple closed rectifiable curve bounding a domain in $\mathbb{C}$, whose closure is contained in $U_{u,x}$. Since the image of Γ by the map $\tau \mapsto u+\tau x$ is compact, there is some $r > 0$ such that $\overline{B(u+\tau x,r)} \subset U$ for all $\tau \in \Gamma$. By Fubini's theorem, formula (II.3.3) and the Cauchy integral theorem yield

$$\int_\Gamma \lambda(\hat{d}^q f(u+\tau x)(y))d\tau =$$

$$= \frac{\|y\|^q}{2\pi i r^q} \int_\Gamma \lambda\left(\int_{|\zeta|=1} \frac{1}{\zeta^{q+1}} f\left(u+\tau x + \frac{\zeta r}{\|y\|} y\right)d\zeta\right)d\tau =$$

$$= \frac{\|y\|^q}{2\pi i r^q} \int_\Gamma \left(\int_{|\zeta|=1} \frac{1}{\zeta^{q+1}} \lambda\left(f\left(u+\tau x + \frac{\zeta r}{\|y\|} y\right)\right)d\zeta\right)d\tau =$$

$$= \frac{\|y\|^q}{2\pi i} \int_{|\zeta|=1} \frac{1}{\zeta^{q+1}} \left(\int_\Gamma \lambda\left(f\left(u+\tau x + \frac{\zeta r}{\|y\|} y\right)\right)d\tau\right)d\zeta = 0,$$

for $\tau \mapsto \lambda\left(f\left(u+\tau x + \frac{\zeta z}{\|y\|} y\right)\right)$ is holomorphic.

By Morera's theorem, the function $\tau \mapsto \lambda(\hat{d}^q f(u+\tau x)(y))$ is holomorphic in $U_{u,x}$.

Q.E.D.

§ 4. Gateaux holomorphy.

We will now prove Theorem II.3.10. We begin by considering the case $E = \mathbb{C}$, proving the following theorem due to N. Dunford.

Let H be a complex normed space, F a complex Banach space, F' the (topological) dual of F, and $F_0' \subset F$ a determining manifold for F.

Theorem II.4.1. *Let* U *be an open subset of* $\mathbb{C}$, *and let* $f: U \to L(H,F)$ *be a function such that, for every* $x \in H$ *and every* $\zeta \in F_0'$, *the complex-valued function* $\zeta \leftrightarrow \lambda(f(\zeta)(x))$ *is holomorphic on* U. *Then* $f \in \mathrm{Hol}(U, L(H,F))$.

We divide the proof into three steps.

1) Lemma II.4.2. *For every* $\zeta_0 \in U$ *the limit*

$$\lim_{\zeta\to 0} \frac{1}{\zeta} (f(\zeta_0 + \zeta) - f(\zeta_0))$$

exists in the norm topology of $L(H,F)$.

Proof. Let Γ be a simple closed rectifiable Jordan curve in $\mathbb{C}$ bounding a domain D such that $\zeta_0 \in D$ and $\overline{D} \subset U$. Let D_0 be a domain for which $\zeta_0 \in D_0 \subset \overline{D_0} \subset D$.

For $\zeta_0 + \zeta_1 \in D_0$, $\zeta_0 + \zeta_2 \in D_0$ the Cauchy integral formula yields

$$\lambda(f(\zeta_0 + \zeta_j)(x)) - \lambda(f(\zeta_0)(x)) =$$

$$= \frac{\zeta_j}{2\pi i} \int_\Gamma \frac{1}{(\tau - \zeta_0)(\tau - \zeta_0 - \zeta_j)} \lambda(f(\tau)(x))\,d\tau \qquad (j=1,2),$$

and therefore, for $0 \neq \zeta_1 \neq \zeta_2 \neq 0$,

$$\frac{1}{\zeta_1 \quad \zeta_2} \{ \frac{1}{\zeta_1} [\lambda(f(\zeta_0 + \zeta_1)(x)) - \lambda(f(\zeta_0)(x))] -$$

$$- \frac{1}{\zeta_2} [\lambda(f(\zeta_0 + \zeta_2)(x)) - \lambda(f(\zeta_0)(x))] \} =$$

$$= \frac{1}{2\pi i} \int_\Gamma \frac{1}{(\tau - \zeta_0)(\tau - \zeta_0 - \zeta_1)(\tau - \zeta_0 - \zeta_2)} \lambda(f(\tau)(x))\,d\tau .$$

The compact set $\overline{D_0}$ has positive distance, r, from Γ. Thus for $\tau \in \Gamma$, $|\tau - \zeta_0| > r$, $|\tau - \zeta_0 - \zeta_1| > r$, $|\tau - \zeta_0 - \zeta_2| > r$, and therefore

$$\sup\{ |\frac{1}{\zeta_1 - \zeta_2} (\frac{1}{\zeta_1} [\lambda(f(\zeta_0 + \zeta_1)(x)) - \lambda(f(\zeta_0)(x))] -$$
$$- \frac{1}{\zeta_2} [\lambda(f(\zeta_0 + \zeta_2)(x)) - \lambda(f(\zeta_0)(x))])| : \zeta_0, \zeta_0 + \zeta_1\ \zeta_0 + \zeta_2 \in D_0\}$$
$$\leqslant \frac{1}{2\pi} \frac{1}{r^3} (\sup\{|\lambda(f(\tau)(x))| : \tau \in \Gamma\}) \int_\Gamma |d\tau| < \infty$$

for every $\lambda \in F_0'$, and every $x \in H$. Keeping x fixed, by the theorem of Banach-Steinhaus there exists a finite positive constant k(x) such that

$$\sup\{ \| \frac{1}{\zeta_1 - \zeta_2} (\frac{1}{\zeta_1} [f(\zeta_0 + \zeta_1)(x) - f(\zeta_0)(x)] -$$
$$- \frac{1}{\zeta} [f(\zeta_0 + \zeta_2)(x) - f(\zeta_0)(x)])\| : \zeta_0, \zeta_0 + \zeta_1, \zeta_0 + \zeta_2 \in D\} \leqslant k(x).$$

Again the theorem of Banach Steinhaus implies that there exists a finite positive constant k such that

$$\sup \| \frac{1}{\zeta_1 - \zeta_2} (\frac{1}{\zeta_1} [f(\zeta_0 + \zeta_0) - f(\zeta_0)]) - \frac{1}{\zeta_2} [f(\zeta_0 + \zeta_2) - f(\zeta_0)])\| :$$
$$\zeta_0,\ \zeta_0 + \zeta_1,\ \zeta_0 + \zeta_2 \in D_0\} \leqslant k.$$

Letting $\zeta_1, \zeta_2 \to 0$, we see that the difference quotients converge to an element of $L(H,F)$. Let $f'(\zeta\ )$ be the limit. Choosing $\zeta_2 = 0$ and letting $\zeta_1 \to 0$ we have

$$\text{(II.4.1)} \qquad \lim_{\zeta_1 \to 0} \| \frac{1}{\zeta_1} (f(\zeta_0 + \zeta_1) - f(\zeta_0)) - f'(\zeta_0)\| = 0.$$

Q.E.D.

<u>Remark</u>. The above theorem shows that (II.4.1) holds uniformly on compact subsets of U.

As a consequence of Lemma II.4.2 we have

Lemma II.4.3. *If* f *satisfies the hypothesis of Theorem* II.4.1, *then* f *is continuous.*

We compute now the integral $\int_\Gamma \frac{1}{\tau - \zeta_0} f(\tau)d\tau$. For every $\lambda \in F_0'$ and every $x \in E$, we have

$$\lambda((\int_\Gamma \frac{1}{\tau - \zeta_0} f(\tau)d\tau)(x)) = \int_\Gamma \frac{1}{\tau - \zeta_0} \lambda(f(\tau)(x))d\tau.$$

2) Since the function $\zeta \mapsto \lambda(f(\zeta)(x))$ is holomorphic on U, by Cauchy's integral formula,

$$\lambda(f(\zeta_0)(x)) = \frac{1}{2\pi i} \int_\Gamma \frac{1}{\tau - \zeta_0} \lambda(f(\tau)(x))d\tau =$$
$$= \lambda(\frac{1}{2\pi i} \int_\Gamma \frac{1}{\tau - \zeta_0} f(\tau)(x)d\tau)$$

for every $\lambda \in F_0'$. Thus

$$f(\zeta_0)(x) = \frac{1}{2\pi i} \int_\Gamma \frac{1}{\tau - \zeta_0} f(\tau)(x)d\tau$$

for every $x \in H$, and therefore

$$f(\zeta_0) = \frac{1}{2\pi i} \int_\Gamma \frac{1}{\tau - \zeta_0} f(\tau)d\tau \qquad (\zeta_0 \in D).$$

3) Let A be the open disc with center ζ_0 and radius equal to the distance of ζ_0 from $\mathbb{C}\setminus D$. Then $A \subset D$, and for every $\zeta \in A$, $\tau \in \Gamma$

$$\frac{1}{\tau - \zeta} = \frac{1}{\tau - \zeta_0 - (\zeta - \zeta_0)} = \frac{1}{\tau - \zeta_0} \frac{1}{1 - \frac{\zeta - \zeta_0}{\tau - \zeta_0}} = \sum_{q=0}^{+\infty} \frac{(\zeta - \zeta_0)^q}{(\tau - \zeta_0)^{q+1}}$$

the convergence being normal for $\tau \in \Gamma$. Hence we can integrate term by term, and letting

$$f_q(\zeta_0) = \frac{1}{2\pi i} \int_\Gamma \frac{1}{(\tau - \zeta_0)^{q+1}} f(\tau)d \qquad (q=0,1,\dots), \tag{II.4.2}$$

we obtain

$$f(\zeta) = \sum_{q=0}^{+\infty} (\zeta - \zeta_0)^q f_q(\zeta_0).$$

The series on the right converges for every $\zeta \in A$. Thus, by Lemma I.2.2, the convergence is uniform on all compact subsets of A. Thus f is holomorphic on A, and that completes the proof of Theorem II.4.1.

Since parts 2) and 3) of the above proof follow from Lemma II.4.2, then we have

<u>Lemma II.4.4.</u> *Let* $f: U \to L(H,F)$ *be a function such that for every* $\zeta_0 \in U$

$$\lim_{\zeta\to 0} \frac{1}{\zeta} (f(\zeta_0 + \zeta) - f(\zeta_0))$$

exists. Then $f \in \text{Hol}(U, L(H,F))$.

The coefficients $d^q f(\zeta)$ of the Taylor expansion of f are given by

$$d^q f(\zeta) = q! f_q(\zeta)$$

where f_q is expressed by (II.4.2). By Proposition II.3.12, which for $E = \mathbb{C}$ follows from Theorem II.4.1, $d^q f: U \to L(H,F)$ is holomorphic.

<u>Lemma II.4.5.</u> *For* $\zeta_0 \in U$

$$d^{q+1} f(\zeta_0) = \lim_{\zeta\to 0} \frac{1}{\zeta} (d^q f(\zeta_0 + \zeta) - d^q f(\zeta_0)).$$

<u>Proof</u>. The existence of the limit on the right follows from Lemma II.4.2. Since for any $\lambda \in F'$ and any $x \in H$

$$\lambda(d^{q+1} f(\zeta_0)(x)) = \lim_{\zeta\to 0} \frac{1}{\zeta} (\lambda(d^q f(\zeta_0 + \zeta)(x)) - \lambda(d^q f(\zeta_0)(x)))$$

the conclusion follows from the Hahn-Banach theorem.

Q.E.D.

The above lemma makes available to $L(H,F)$-valued holomorphic functions of one complex variable the standard procedures of calculus. We mention for future reference the 'mean value theorem':

<u>Theorem II.4.6</u>. *Let* U *be open and convex in* $\mathbb{C}$, *and let* $f \in \mathrm{Hol}(U, L(H,F))$. *Then for* $\zeta_1, \zeta_2 \in U$, $\zeta_1 \neq \zeta_2$

$$\left\|\frac{1}{\zeta_2 \zeta_1}(f(\zeta_2) - f(\zeta_1))\right\| \leqslant \sup \|d^1 f(\zeta_1 + t(\zeta_2 - \zeta_1))\| : 0 \leqslant t \leqslant 1\}.$$

For a proof, cf. e.g. [Dieudonné, 1, (8.5.1)-(8.5.4), pp. 160-162].

We come now to the proof of Theorem II.3.10. With H, F, F', F_0' as before, U is now an open subset of a complex normed space E. For $u \in U$, $v \in E$ we denote, as before, by $U_{u,v}$ the open subset of $\mathbb{C}$: $U_{u,v} = \{\zeta \in \mathbb{C}: u + \zeta v \in U\}$.

<u>Theorem II.4.7</u>. *Let* $f: U \to L(H,F)$ *be a locally bounded function such that for every choice of* $u \in U$, $v \in E$, $z \in H$, $\lambda \in F_0'$, *the complex valued function* $\zeta \mapsto \lambda(f(u+\zeta v)(z))$ *is holomorphic on* $U_{u,v}$. *Then* $f \in \mathrm{Hol}(U, L(H,F))$.

We divide the proof into four steps.

a) The function on $U_{u,v}$

$$\zeta \mapsto f(u + \zeta v)$$

satisfies the hypotheses of Theorem II.4.1. Hence it is holomorphic; in particular it is continuous, and therefore can be integrated. More specifically, let $r > 0$ be such that $\overline{B(u,r)} \subset U$ and that f is bounded by a finite constant $M > 0$. For any $v \in E \setminus \{0\}$ with $\|v\| < r$, the closed unit disc in $\mathbb{C}$ is contained in $U_{u,v}$. We define $P_q(v)$ by the integral

$$\text{(II.4.3)} \qquad P_q(v) = \frac{\|y\|^q}{2\pi i} \int_0^{2\pi} e^{-iq\theta} f\left(u + \frac{e^{i\theta}}{\|v\|} v\right) d\theta \qquad (q = 0, 1, \dots)$$

We will prove the theorem by showing that $v \mapsto P_q(v)$ is a continuous q-homogeneous polynomial from E to $L(H,F)$, and that

$$f(u+v) = \sum_{q=0}^{+\infty} P_q(v)$$

uniformly for $\|v\| \leqslant s$ whenever $0 < s < r$. Since by (II.4.3)

(II.4.4) $$\|P_q(v)\| \leqslant M \quad \text{for all} \quad v \in B(0,r),$$

then we need only show that P_q is a q-homogeneous polynomial from E to $L(H,F)$.

b) We consider first the case in which $E = \mathbb{C}^n$, for some $n \geqslant 1$, and $F = H = \mathbb{C}$. By a theorem of F. Hartogs (which does not require local boundedness, cf. e.g. [Hörmander, 1; Theorem 2.2.8, p. 28]) f is holomorphic on U. Let

$$f(u+v) = \sum_{q=0}^{+\infty} Q_q(v)$$

be the power series expansion of f at u. Here Q_q is a q-homogeneous (continuous) polynomial and the convergence is uniform on $B(u,r')$ for some $r' > 0$. Integrating term by term we obtain

$$Q_q(v) = \frac{\|y\|^q}{2\pi i} \int_0^{2\pi} e^{-iq\theta} f\left(u + \frac{e^{i\theta}}{\|v\|} v\right) d\theta = P_q(v)$$

for all v with $\|v\| < \min(r,r')$. Thus $P_q = Q_q$ and that proves Theorem II.4.7 in the particular case $E = \mathbb{C}^n$, $H = F = \mathbb{C}$.

c) Going back to the general case, we define a function $A_q: E^q \to L(H,F)$ by putting $\hat{A} = P$ in the polarization formula of Lemma I.1.1. Let E_0 be a finite-dimensional complex subspace of E. By b) the restriction of $\lambda(A_q(\cdot)(z))$ is a q-linear map of E_0 into $\mathbb{C}$. Since E_0 and λ are arbitrary, $A_q(\cdot)(z)$ is a q-linear map of E^q into F. Thus $A_q(\cdot)$ is a q-linear map of E^q into $L(H,F)$, and therefore P_q is a q-homogeneous polynomial from E to $L(H,F)$. By the final remark in a) $P_q \in P^q(E, L(H,F))$.

d) Let $v \in E$ with $\|v\| \leqslant r' < r$. By Theorem II.4.1

$$f(u+\zeta v) = \sum_{q=0}^{+\infty} P_q(\zeta v)$$

for all $|\zeta| \leqslant 1$, and therefore

$$f(u+v) = \sum_{q=0}^{+\infty} P_q(v)$$

for all $v \in B(0,r)$. By (II.4.4)

$$\|P_q(v)\| = \|P_q(\frac{rv}{r})\| = \frac{1}{r^q}\|P_q(rv)\| \leqslant \frac{M}{r^q}$$

for all $v \in B(0,1)$, then

$$\|P_q\| \leqslant \frac{M}{r^q},$$

and therefore

$$\limsup \|P_q\|^{\frac{1}{q}} \leqslant \frac{1}{r}.$$

By Lemma I.2.1 the series $\sum_{q=0}^{+\infty} P_q(v)$ converges uniformly for $v \in B(0,r')$ for all $0 < r' < r$. That completes the proof of Theorem II.4.7.

Remark 1. Local boundedness is not required in Theorem II.4.1, and enters the proof of Theorem II.4.7 only to establish inequality (II.4.4), which in part d) yields the uniform convergence of $\sum_{q=0}^{+\infty} P_q(v)$. The following theorem collects the results obtained so far, which do not involve local boundedness. For any $u \in U$, let $\Omega(u)$ be the largest open balanced neighborhood of 0 such that $u + \Omega(u) \subset U$ (we recall that a subset $K \subset E$ is balanced, or complete circular, if $x \in K$ and $|\zeta| \leqslant 1$ implies $\zeta x \in K$).

Theorem II.4.8. *Let* $f: U \to L(H,F)$ *be a function such that for every choice of* $u \in U$, $v \in E$, $\lambda \in F'_0$ *the complex-valued function* $\zeta \mapsto \lambda(f(u+\zeta v))$ *is holomorphic on* $U_{u,v}$.

Then for every $u \in U$ *there exists a unique sequence* $(P_q)_q$ *of* q-*homogeneous polynomials from* E *to* $L(H,F)$ *such that*

$$f(u+x) = \sum_{q=0}^{+\infty} P_q(x) \quad \textit{for all} \quad x \in \Omega(u).$$

Proof. Exercise. (*Hint:* Apply Theorem II.4.1 to the function $f|_{U_{u,v}}$ to define P_q by (II.4.3). Then apply steps b) and c) in the proof of Theorem II.4.7.

Functions satisfying the hypothesis of Theorem II.4.8 are called *Gateaux holomorphic*. Theorem II.4.7 and Remark 2 yield

Corollary II.4.9. *For Gateaux holomorphic functions in a Banach space local boundedness is equivalent to continuity.*

We will now discuss briefly different kinds of convergence of power series expansions. This investigation will be pursued further in § 5. We begin with an example.

Examples. 1. As usual c_0 will denote the Banach space of all sequences $x = (\zeta^1, \zeta^2, \dots)$ of complex numbers, converging to zero, with norm $\|x\| = \sup|\zeta^\nu|$. Let P_q be the q-homogeneous polynomial from c_0 to $\mathbb{C}$ defined by

$$P_q(x) = \zeta^1 \dots \zeta^q \qquad \text{if} \quad q > 0,$$

and by $P_q(x) = 1$ if $q = 0$. Then $P_q \in P^q(c_0, \mathbb{C})$, and $\|P_q\| = 1$.

Thus the power series $\sum_{q=0}^{+\infty} P_q$ has radius of convergence $R = 1$.

For any $x = (\zeta^1, \zeta^2, \dots) \in c_0$ and any $\nu = 1, 2, \dots$, let $x^{(\nu)} = (\zeta^\nu, \zeta^{\nu+1}, \dots)$. Then $x^{(\nu)} \in c_0$, and $\|x^{(\nu)}\| \leqslant \|x\|$. For any ε such that $0 < \varepsilon < 1$ there is an index q_0 such that

$|\zeta^q| < \varepsilon$ whenever $q > q_0$. Thus for $q > q_0$

$$1+|\zeta^q|+|\zeta^q||\zeta^{q+1}|+|\zeta^q||\zeta^{q+1}||\zeta^{q+2}|+\ldots < 1+\varepsilon+\varepsilon^2+\ldots = \frac{1}{1-\varepsilon} .$$

Therefore the series $1+\zeta^q+\zeta^q\zeta^{q+1}+\ldots$ is absolutely convergent for $q = 1,2,\ldots$, and

$$1+|\zeta^q|+|\zeta^q||\zeta^{q+1}|+\ldots \leqslant 1+\|x\|+\ldots+\|x\|^{q_0-1} + \frac{\|x\|^{q_0}}{1-\varepsilon}$$

Setting

$$\text{(II.4.5)} \qquad f(x) = 1 + \zeta^1 + \zeta^1\zeta^2 + \ldots = \sum_{q=0}^{+\infty} P_q(x)$$

we obtain a map $f\colon c_0 \to \mathbb{C}$ for which

$$\text{(II.4.6)} \qquad |f(x^{(\nu)})| \leqslant 1 + \|x\| + \ldots + \|x\|^{q_0-1} + \frac{\|x\|^{q_0}}{1-\varepsilon} \qquad (\nu=1,2,\ldots).$$

For any $y = (\eta^1,\eta^2,\ldots) \in c_0$, with $\|y\| < 1$, we have

$$f(x+y) = 1 + (\zeta^1+\eta^1) + (\zeta^1+\eta^1)(\zeta^2+\eta^2) + \ldots =$$

$$= f(x) + f(x^{(2)})\eta^1 + f(x^{(3)})\eta^1\eta^2 + \ldots = \sum_{q=0}^{+\infty} Q_q(y),$$

where Q_q is the q-homogeneous polynomial from c_0 to $\mathbb{C}$ defined by

$$Q_q(y) = f(x^{(q+1)})\eta^1\ldots\eta^q = f(x^{(q+1)})P_q(y).$$

Thus $\|Q_q\| = |f(x^{(q+1)})|$, and therefore, by (II.4.6),

$$\limsup \|Q_q\|^{\frac{1}{q}} \leqslant 1,$$

i.e. the radius of convergence of the power series $\sum_{q=0}^{+\infty} Q_q$ is $\geqslant 1$. Hence $f \in \mathrm{Hol}(c_0,\mathbb{C})$. Note that, although f is holomorphic on the entire space c_0, its local representation requires several power series expansions.

Since the radius of convergence of $\sum_{q=0}^{+\infty} P$ is equal to 1, $\sum_{q=0}^{+\infty} P_q(x)$ converges for every $x \in c_0$ but is normally convergent only for $\|x\| \leqslant r$ for any $r < 1$.

Furthermore, for $\|x\| < 1$, (II.4.5) yields

$$|f(x)| \leqslant 1 + \|x\| + \|x\|^2 + \dots = \frac{1}{1-\|x\|},$$

while for $x = (\underbrace{1,\dots,1}_{q},0,\dots)$, $f(x) = q+1$.

Hence f is bounded on any ball $B(0,r)$ with center 0 and radius $r<1$, and is unbounded on any ball $B(0,r)$ with $r>1$. The fact that the separating value 1 is the radius of convergence of $\sum_{q=0}^{+\infty} P_q$ is not casual, as we shall see.

2. Let $E = F = \ell_2(\mathbb{N})$ and let $P_q \in \mathcal{P}^q(\ell_2(\mathbb{N}),\ell_2(\mathbb{N}))$ be the polynomial defined by

$$P_q(x) = (0,0,\dots,0,\zeta_q^q,0,\dots) \qquad (q = 0,1,\dots)$$

for $x = (\zeta_0,\zeta_1,\dots,\zeta_q,\dots) \in \ell_2(\mathbb{N})$.

Consider the power series $\sum_{q=0}^{+\infty} P_q$; its radius of convergence is 1, but the series converges at each point of $\ell_2(\mathbb{N})$.

However, if we consider $E = \ell_\infty(\mathbb{N})$, $F = \ell_\infty(\mathbb{N})$, and define P_q in the same way, the series $\sum_{q=0}^{\infty} P_q$ has the same radius of convergence 1, but the largest domain on which the series is (pointwise) convergent is the set

$$\{x \in \ell_\infty(\mathbb{N}) : \limsup_{q\to\infty} |\zeta_q| < 1\}$$

for $x = (\zeta_0,\zeta_1,\dots,\zeta_q,\dots)$.

For any $f \in \mathrm{Hol}(U,F)$ and for any $u \in U$ the *radius of boundedness* of f at u is, by definition, the largest r such that $B(u,r) \subset U$, and f is bounded on $\overline{B(u,s)}$ with $s < r$.

<u>Proposition II.4.10</u>. *Let* r_b *and* R *be the radius of boundedness at* u *and the radius of convergence of the Taylor expansion of* f *at* $u \in U$, *and let* d *be the distance of* u *from* $E\setminus U$.

Then

$$r_b = \inf (R,d).$$

<u>Proof</u>. By definition $r_b \leqslant d$. For $0 < s < r_b$, let

$$M_s = \sup \{\| f(u+x) : \|y\| \leqslant s\}.$$

Then, by the Cauchy inequalities,

$$\|\frac{1}{q!} \hat{d}^q f(u)\| \leqslant \frac{M_s}{s^q} \qquad (q=0,1,2,\dots).$$

By Lemma I.2.1, one has $R \geqslant s$ whenever $0 < s < r_p$, and therefore $R \geqslant r_b$. To prove that $r_b \geqslant \inf(R,d)$ choose any s with $0 < s < \inf(R,d)$. The power series expansion of f at u converges uniformly on $\overline{B(u,s)}$. Letting $P_q(y) = \frac{1}{q!} \hat{d}^q f(u)(y)$ by definition of convergence, we have

$$\| f(u+x)\| \leqslant \sum_{q=0}^{+\infty} \|P_q(x)\| \leqslant \sum_{q=0}^{+\infty} \|P_q\| s^q < \infty$$

for all $x \in \overline{B(0,s)}$. Hence $s \leqslant r_b$, and therefore $r_b \geqslant \inf(R,d)$.

Q.E.D.

<u>Remarks</u>. 3. In the above example $d = \infty$, and $r_b = R = 1$ at 0.

4. If E has finite dimension, every ball with finite radius has compact closure. Hence every $f \in \mathrm{Hol}(U,F)$ is bounded in any closed ball contained in U. Thus, if $\dim_{\mathbb{C}} E < \infty$, $r_b = d$ for all $u \in U$ and any $f \in \mathrm{Hol}(U,F)$. The above example shows that this is not always the case when E has infinite dimension.

Given $f \in \mathrm{Hol}(U,F)$ we denote by $r_b(u)$ the radius of boundedness of f at $u \in U$. For any r, $0 < r < r_b(u)$, let s be such that $r < s < r_b(u)$, and let $\delta = r_b(u) - s$.

If $\|v-u\| \leqslant \delta$, then, for any $x \in \overline{B(v,s)}$,

$$\|x-u\| \leqslant \|x-v\| + \|v-u\| < s + \delta = r_b(u),$$

i.e.

$$\|v-u\| < \delta \Rightarrow \overline{B(v,s)} \subset B(u,r_b(u)).$$

Thus, by Proposition II.4.10, $B(u,s) \subset U$, and $r_b(v) > r$. That proves

<u>Lemma II.4.11</u>. *For any* $f \in \mathrm{Hol}(U,F)$ *the function* $u \mapsto r_b(u)$ *is lower semi-continuous.*

Thus $u \mapsto r_b(u)$ attains its (positive) minimum value on every compact subset of U. Proposition II.4.10 yields then

<u>Lemma II.4.12</u>. *For every* $f \in \mathrm{Hol}(U,F)$ *and for every compact subset* $K \subset U$ *there exists a positive constant* c *which is smaller than the radius of convergence* $R(u)$ *of the Taylor expansion of* f *at any point* $u \in K$.

§ 5. The Zorn theorem.

Let E and F be complex Banach spaces. The following lemma is more precise than Theorem I.1.4.

<u>Lemma II.5.1</u>. *Let* P_q *be a* q-*homogeneous polynomial from* E *to* F. *If there is* $x_0 \in E$, *a balanced subset* $K \subset E$ *and a finite* $M > 0$ *such that*

$$\|P_q(x)\| \leqslant M,$$

for all $x \in x_0 + K$, *then*

$$\|P_q(x)\| \leqslant M \quad \textit{for all} \quad x \in \overline{\Delta}\zeta_0 + K.$$

<u>Proof</u>. For $y \in K$, the map $\zeta \mapsto P_q(\zeta x_0 + y)$ of $\mathbb{C}$ into F is holomorphic. By the maximum principle, its restriction to $\overline{\Delta}$ takes its maximum (for the norm) at some point $\zeta = e^{i\theta}$. The polynomial P_q being homogeneous, and K being circular

$$\|P_q(e^{i\theta}x_0+y)\| = \|P_q(x_0+e^{-i\theta}y)\| \leqslant \sup\{\|P_q(x_0+x)\| : x \in K\} \leqslant M.$$

Q.E.D.

<u>Corollary II.5.2</u>. *Let* $(P_q)_{q=0,1,2,\dots}$ *be a sequence of* q-*homogeneous polynomials. Let* $x_0 \in E$ *and let* A *be an open balanced neighborhood of* 0, *such that*

$$\sup\{\|P_q(x)\| : x \in x_0 + A\} \leqslant M \qquad (q=0,1,2,\dots)$$

for some $M > 0$. *Then*

$$\sup\{\|P_q(x)\| : x \in \overline{\Delta}x_0 + A\} \leqslant M \qquad (q=0,1,2,\dots).$$

In particular

$$\sup\{\|P_q(x)\| : x \in A\} \leqslant M.$$

<u>Proposition II.5.3</u>. *Let* E *and* F *be complex Banach spaces. Let* D *be an open balanced neighborhood of* 0 *in* E, *and let* $(P_q)_{q=0,1,\dots}$ *be a sequence of* q-*homogeneous polynomials* $P_q \in P^q(E,F)$ *from* E *to* F, *such that the series* $\sum_{q=0}^{+\infty} P_q(x)$ *converges to* f(x) *at each point* $x \in D$. *Then* $f \in \mathrm{Hol}(D,F)$.

<u>Proof</u>. a) We prove first that f is holomorphic in an open neighborhood of 0.

The domain D, as an open subset of the Banach space E, is a Baire space(★). Since the functions P_q are continuous, and $\sum_{q=0}^{+\infty} P_q(x)$ converges at each point $x \in D$, there exists a non-empty open subset U of D and a finite constant $M > 0$ such that

$$\|P_0 + P_1(x) + \dots + P_q(x)\| \leqslant M \quad \text{for all} \quad x \in U \quad \text{and} \quad q=0,1,2,\dots .$$

Hence

$$\|P_q(x)\| \leqslant 2M \quad \text{for all} \quad x \in U \quad \text{and} \quad q=0,1,2,\dots .$$

(★) All the basic notions on Baire spaces can be found in Appendix A.

By Corollary II.5.2 there is an open ball $B(0,r)$ with $r > 0$ such that

$$\|P_q(x)\| \leqslant 2M \quad \text{for all} \quad x \in B(0,r), \quad q=0,1,\dots .$$

Hence

$$\|P_q(x)\| = \frac{1}{r^q}\|P_q(rx)\| \leqslant \frac{2M}{r^q} \quad \text{for } \|x\| < 1, \qquad q=0,1,\dots .$$

and therefore

$$\|P_q\| \leqslant \frac{2M}{r^q} \qquad \text{for} \quad q=0,1,\dots .$$

Thus

$$\limsup \|P_q\|^{\frac{1}{q}} \leqslant \frac{1}{r},$$

i.e. the radius of convergence of the power series $\sum_{q=0}^{+\infty} P_q$ is $\geqslant r$. By consequence the series $\sum_{q=0}^{+\infty} P_q(x)$ converges uniformly for $x \in B(0,s)$ for any $0<s<r$. Thus $f_{|B(0,r)} \in \mathrm{Hol}(B(0,r),F)$.

b) In view of a), the theorem will be completely proved once we show that for any $u \in D$ there is a sequence $(Q_q)_{q=0,1,\dots}$, with $Q_q \in P^q(E,F)$ such that

$$f(u+y) = \sum_{q=0}^{+\infty} Q_q(y)$$

for all y in a neighborhood of 0. Let $r > 0$ be such that $B(u,3r) \subset D$, and let $\|y\| \leqslant r$.

Let $\varepsilon > 0$ be such that $\tau u+B(0,2r) \subset D$ for all τ with $|\tau| < 1+\varepsilon$. Then for $|\tau| < 1+\varepsilon$, $|\zeta| \leqslant 2$,

$$f(\tau u+\zeta y) = \sum_{q=0}^{+\infty} P_q(\tau u+\zeta y) = \sum_{q=0}^{+\infty} \sum_{\substack{p_1+p_2=q \\ p_1 \geq 0,\, p_2 \geq 0}} \tau^{p_1}\zeta^{p_2} a_{p_1 p_2} .$$

By Abel's lemma (Lemma I.2.2) the convergence is normal on compact subsets of the bi-disc $\{|\zeta|<1+\varepsilon,\ |\tau|<2\} \subset \mathbb{C}^2$. Thus we can permute the terms on the right hand side, and for $|\zeta| \leqslant 1$ we have

$$\text{(II.5.1)} \qquad f(u+\zeta y) = \sum_{q=0}^{+\infty} \zeta^q Q_q(y)$$

where Q_q is a q-homogeneous polynomial from E to F.

c) Comparing with

$$\text{(II.5.2)} \qquad f(u+\zeta y) = \sum_{q=0}^{+\infty} P_q(u+\zeta y)$$

and taking into account normal convergence on a neighborhood of $\bar{\Delta}$ in (II.5.1) and (II.5.2) - which allows for differentiation term by term - we have

$$q!Q_q(y) = \lim_{p\to\infty} \sum_{0}^{p} \left(\frac{d^q}{d\zeta^q} P_q(u+\zeta y) \right)\Big|_{\zeta=0} \qquad (y \in B(0,r)).$$

Thus, by Theorem I.1.9, $Q_q \in P^q(E,F)$.

Q.E.D.

Theorem II.5.4. *(M.A. Zorn) Let* U *be an open set in* E*, and let* $f: U \to F$ *be Gateaux holomorphic on* U*. The set* U' *of points of* U *at which* f *is continuous is open and closed in* U.

Proof. For any $u \in U$, $\Omega(u)$ will indicate the largest open balanced neighborhood of $0 \in E$ such that $u + \Omega(u) \subset U$.

a) We will show that, if $u \in U'$, then $u + \Omega(u) \subset U'$. (Note that this fact will imply incidentally that U' is open, which is obvious). Let λ be a continuous linear form on F. By Theorem II.4.8 there exists a unique sequence of q-homogeneous polynomials P_q from E to F such that

$$f(u+x) = \sum_{q=0}^{+\infty} P_q(x)$$

for all $x \in \Omega(u)$. Since $u \in U'$ there is also a sequence of *continuous* q-homogeneous polynomials R_q from E to F, such that

$$f(u+x) = \sum_{q=0}^{+\infty} R_q(x)$$

for all $x \in V$, V being an open neighborhood of 0 in E.

For any continuous linear form λ on F and every ζ in a neighborhood of 0 in $\mathbb{C}$,

$$\sum_{q=0}^{+\infty} \lambda(R_q(\zeta x)) = \sum_{q=0}^{+\infty} \lambda(P_q(\zeta x)).$$

By the identity principle for complex-valued power series in one variable we have

$$\lambda(R_q(x)) = \lambda(P_q(x)) \qquad (q=0,1,2,\ldots)$$

for every continuous linear form λ on F. Hence $P_q = R_q \in \in P^q(E,F)$. By Proposition II.5.3, $f \in \mathrm{Hol}(u+\Omega(u))$, and therefore $u+\Omega(u) \subset U'$.

b) Let $U'' = U \setminus U'$. Suppose that $U'' \neq \emptyset$, and let $v \in U''$. Since the function $u \mapsto \mathrm{dist}(u, E \setminus U)$ is continuous, the set of points $u \in U$ such that $v \in u+\Omega(u)$ contains a neighborhood V of v in U. Let $u \in V$. If $u \in U'$, then by a) $u+\Omega(u) \subset \subset U'$. Therefore $v \in V \subset U'$. This contradiction proves that $V \subset U''$. Thus U'' is open.

Q.E.D.

Corollary II.5.5. *Let* D *be a domain in* E*, and let* $f\colon D \to F$ *be a Gateaux holomorphic function. If* f *is locally bounded at some point of* D*, then* $f \in \mathrm{Hol}(D,E)$.

The following example [Noverraz, 2, p. 32] shows that the Zorn theorem is not always true if E is not complete.

Example. The set c_{00} of all sequences of complex numbers having only a finite number of non vanishing terms is a dense linear subvariety of the Banach space c_0. Let $P_q : c_{00} \to \mathbb{C}$ be the polynomial

$$P_q(x) = (\zeta^1)^{q-1}\zeta^q$$

where $x = (\zeta^1, \zeta^2, \dots) \in c_{00}$ and $q = 1, 2, \dots$. Then $P_q \in \mathcal{P}^q(c_{00}, \mathbb{C})$, being in fact $\|P_q\| = 1$. For every $x \in c_{00}$ the power series $\sum_{q=1}^{+\infty} P_q$ contains only a finite number of non vanishing terms. Thus we define $f : c_{00} \to \mathbb{C}$ by $f(x) = \sum_{q=1}^{+\infty} P_q(x)$. Clearly f is Gateaux holomorphic. Let U be the domain in c

$$U = \{x = (\zeta^1, \zeta^2, \dots) \in c_{00} : |\zeta^1| < 1\}.$$

Let $x_0 = (\zeta^1, \dots, \zeta_0^N, 0, 0, \dots) \in U$, and let $0 < r < 1$ be such that $B(x_0, 2r) \subset U$. Then for any $x = (\zeta^1, \zeta^2, \dots) \in B(x_0, r)$ we have $|\zeta^1| < |\zeta_0^1| \;\; r < 1$, $|\zeta_1^j| < |\zeta_0^j| + r$ for $j = 2, \dots, N$, $|\zeta^j| < r$ for $j > N$, and therefore

$$|f(x)| \leqslant |\zeta^1| + |\zeta^1||\zeta^2| + |\zeta^1|^2|\zeta^3| + \dots + |\zeta^1|^{N+1}|\zeta^N| + \\ + |\zeta^1|^{N+2}|\zeta^{N+1}| + \dots \\ \leqslant (|\zeta_0^1| + r)(1 + |\zeta_0^2| + r + (|\zeta_0^1| + r)(|\zeta_0^3| + r|) + \dots \\ + (|\zeta_0^1| + r)^N(|\zeta_0^N| + r)) + \frac{r(|\zeta_0^1| + r)^{N+2}}{1 - (|\zeta_0^1| + r)} .$$

Thus f is bounded on $B(x_0, r)$, i.e. f is locally bounded on U. On the other hand f is not locally bounded on c_{00}. In fact, let $x = (1, 0, \dots)$ and, for any $\nu = 1, 2, \dots$, let

$$y_\nu = (\underset{1}{\frac{1}{\sqrt{\nu}}}, 0, \dots, \underset{\nu+1}{\frac{1}{\sqrt{\nu}}}, 0, \dots).$$

Then $\|y_\nu\| = \dfrac{1}{\sqrt{\nu}}$, and

$$f(x + y_\nu) = 1 + \frac{1}{\sqrt{\nu}} + \dots + (1 + \frac{1}{\sqrt{\nu}})^\nu \frac{1}{\sqrt{\nu}} .$$

Since $f(x) = 1$, while

$$\lim_{\nu\to\infty} (x + y_\nu) = x \qquad \lim_{\nu\to\infty} f(x + y_\nu) = +\infty ,$$

there is no neighborhood of x on which f is bounded.

§ 6. Plurisubharmonic and plurisuperharmonic functions.

Let E be a complex normed space and let D be an open set in E. A function $\phi: D \to [-\infty, +\infty)$ is said to be *pluri-subharmonic* on D if ϕ is upper semi-continuous on D and if, for every choice of x,y in E, the function $\zeta \mapsto \phi(u + \zeta y)$ is subharmonic on the open set $U_{x,y} = \{\zeta \in \mathbb{C}: x + \zeta y \in D\} \subset \mathbb{C}$. For the theory of subharmonic functions cf. e.g. [Hörmander, § 1.6] or [Radò, 1] .

A function $\phi: D \to (-\infty, +\infty]$ is said to be *plurisuper-harmonic* on D if $-\phi$ is plurisubharmonic on D.

The following lemma is an immediate consequence of the definition and of the corresponding property of subharmonic functions.

Lemma II.6.1. *The upper envelope of a family of plurisubharmonic functions on* D *is plurisubharmonic provided that it is upper semi-continuous and less than* $+\infty$ *everywhere.*

Let F be a complex Banach space and let $f \in Hol(U,F)$. For every choice of x,y in E and of $\alpha \in \mathbb{C}$, the function $\varphi: \zeta \mapsto \exp(\alpha\zeta) f(x + \zeta y)$ is holomorphic in the open set $U_{x,y} \in \mathbb{C}$.

Let $\zeta_0 \in U_{x,y}$ and let $u_0 = x_0 + \zeta_0 y$. We can let $r > 0$ be such that the disc with center ζ_0 and radius $2r$ in $\mathbb{C}$ is contained in $U_{x,y}$. By the Cauchy formula

$$\exp(\alpha\zeta_0)f(u_0) = \frac{1}{2\pi}\int_0^{2\pi} \exp(\alpha(\zeta_0+re^{i\theta}))f(u_0+re^{i\theta}y)\,d\theta$$

and therefore

$$|\exp(\alpha\zeta_0)|\,\|f(u_0)\| \leqslant \frac{1}{2\pi}\int_0^{2\pi} |\exp(\alpha(\zeta_0+re^{i\theta}))|\,\|f(u_0+re^{i\theta}y)\|\,d\theta .$$

Thus the function $\zeta \mapsto \exp(\alpha\zeta)|\,\|f(x+\zeta y)\|$ is subharmonic on $U_{x,y}$, for every choice of $\alpha \in \mathbb{C}$. This implies that the function $\zeta \mapsto \log\|f(x+\zeta y)\|$ is subharmonic on $U_{x,y}$ [Radò, 1; p.15] and in conclusion we have

<u>Lemma II.6.2</u>. *If* $f \in \mathrm{Hol}(U,F)$, *(*F *Banach space), the function* $u \mapsto \log\|(f(u))\|$ *is plurisubharmonic on* U.

<u>Exercise</u>. Let E be a Banach space, let U be an open set in E and let ϕ be a plurisubharmonic function on U. If V is an open set in a complex normed space G, and if $h \in \mathrm{Hol}(V,E)$, then $\phi\circ h$ is a plurisubharmonic function on the open set $V \cap \cap h^{-1}(U)$.

Let F be a complex Banach space and let $f \in \mathrm{Hol}(U,F)$. For any $u \in U$, the Taylor expansion of f at u

$$f(x) = \sum_{q=0}^{+\infty} \frac{1}{q!}\hat{d}^q f(u)(x-u) \tag{II.6.1}$$

has a positive radius of convergence $R(u)$.

Let $\log R$ be the function $u \mapsto \log R(u)$. We shall prove

<u>Theorem II.6.3</u>. *The function* $-\log R$ *is plurisubharmonic on* U.

<u>Proof</u>. By Proposition II.3.11 the function f defined by (II.6.1) for $x \in B(u,R(u))$ is holomorphic. Thus, by Proposition II.4.10 the radius of boundedness $r_b(x)$ of f at $x \in \in B(u,R(u))$ is given by

$$r_b(x) = \inf(R(x),\ \mathrm{dist}(x,\ E\setminus B(u)))).$$

Since

$$\mathrm{dist}(x,\ E\setminus B(u,R(u))) \geqslant R(u) - \|x-u\|,$$

then

$$R(x) \geqslant r_b(x) \geqslant R(u) - \|x-u\| \qquad (x \in B(u,R(u))).$$

That proves that (R and therefore also) $\log R$ is lower semicontinuous.

Let $0 < r < R(u)$ and let $x \in B(u, \frac{r}{2})$. Then by Proposition II.3.6

$$\frac{1}{q!}\|\hat{d}^q f(x)\| \leqslant (\tfrac{r}{2})^{-q}\sup\{\|f(y)\| : y \in B(x,\tfrac{r}{2})\} \leqslant (\tfrac{r}{2})^{-q}\sup\{\|f(y)\| : y \in B(u,r)\}.$$

Since $y \mapsto f(y)$ is bounded on $B(u,r)$, then the sequence $((\frac{1}{q!}\|\hat{d}^q f(x)\|)^{\frac{1}{q}})_{q=1,2,\dots}$ is uniformly bounded for $x \in B(u, \frac{r}{2})$.

Let $x \in B(u,\frac{r}{2})$. The open disc $\{\zeta \in \mathbb{C} : u + \zeta x \in B(u,\frac{r}{2})\} \subset \mathbb{C}$ contains the closure $\bar{\Delta}$ of the unit disc. Since the function $\hat{d}^q f$ is holomorphic (Proposition II.3.12), and since the function $\log\|\hat{d}^q f\|$ is therefore plurisubharmonic, then

$$\log \|\hat{d}^q f(u)\| \leqslant \frac{1}{2\pi}\int_0^{2\pi} \log\|\hat{d}^q f(u + e^{i\theta}z)\|\,d\theta.$$

Hence by Fatou's lemma

$$\log\frac{1}{R(u)} = \limsup \frac{1}{q}\log\frac{1}{q!}\|\hat{d}^q f(u)\| \leqslant$$

$$\leqslant \limsup \frac{1}{q}\frac{1}{2\pi}\int_0^{\pi} \log\frac{1}{q!}\|\hat{d}^q f(u+e^{i\theta}x)\|\,d\theta$$

$$\leqslant \frac{1}{2\pi}\int_0^{2\pi} \limsup \frac{1}{q}\log\frac{1}{q!}\|\hat{d}^q f(u+e^{i\theta}x)\|\,d\theta =$$

$$= \frac{1}{2\pi}\int_0^{2\pi}\log\frac{1}{R(u+e^{i\theta}x)}\,d\theta.$$

That proves that the function $-\log R$ is plurisubharmonic

on U.

Q.E.D.

Corollary II.6.4. *The function* $u \mapsto R(u)^{-1}$ *is plurisubharmonic on* U.

Notes.

The definition of a Banach valued holomorphic function defined on a domain in a complex Banach space is, e.g., in [Douady, 1] - together with examples 1, 2 and 3 of § 1 - and in [Nachbin, 1] .

The proof of the inverse mapping theorem in § 2 follows closely the proof given in [Douady, 1] and is inspired by [Cartan, 1] . Some of the results of § 3 concerning the Taylor expansion, the Cauchy inequalities and their consequences appear in [Nachbin, 1]. The fact - proved in § 3 (Proposition II.3.12) - that the terms of the Taylor expansion around a point are holomorphic functions of the latter are instrumental in establishing the logarithmic superharmonicity of the radius of convergence (Theorem II.6.3). All these results are obtained by means of the characterization of Banach valued holomorphic functions in terms of Gateaux analyticity and local boundedness. In case of Banach valued holomorphic functions of one complex variable, local boundedness turns out to be redundant, as it is shown by Theorem II.4.1 (Dunford's theorem). The proof of this theorem, given in § 4, follows closely the one appearing in [Hille-Phillips, 1, pp.93-94] ; cf. also [Yosida, 1, pp. 128-129] . Passing to functions defined on a domain in a complex Banach space, local boundedness plays a crucial role in the proof of Theorem II.4.7; here the proof uses Theorem II.4.1 and then follows the ideas of the proof given in [Douady, 1]. For Gateaux holomorphic functions, continuity is equivalent to local boundedness (Corollary II.4.9) and characterizes holomorphic functions. This is the content of the Zorn theorem (Theorem II. 5.4), whose proof - as it appears in § 5 - follows essentially the original proof given by M.A. Zorn in [Zorn, 1 and 2]; cf. also [Hille-Phillips, 1]. The related notion of radius of boundedness and the discussion of different kinds of convergence of power series follows [Nachbin, 1]. The example in § 5, illustrating the role played by completeness is taken from [Noverraz, 1]. We refer the reader to [Noverraz, 1 and 2] for a discussion of holomorphy and Gateaux holomorphy for mappings from a domain in a locally convex space into a locally convex space.

For a discussion of the Fréchet differentiability of holomorphic functions cf. e.g. [Hille-Phillips, 1] and [Dieudonné, 1].

CHAPTER III

MAXIMUM PRINCIPLES

A maximum principle for vector valued holomorphic functions was established in § II.3, which turns out to be weaker than the classical maximum principle. A stronger form will be discussed in this chapter. A "Schwarz lemma" due to L.A.Harris will be established in § III.2.

§ 1. A strong maximum principle.

According to the maximum principle for scalar valued holomorphic functions, the presence of a local maximum on a domain in $\mathbb{C}$ implies that the function itself is constant, and not only its maximum. The following example shows that such a maximum principle does not always hold for vector valued holomorphic functions. Let $F = \mathbb{C}^2$ with norm $\|(\zeta^1,\zeta^2)\| = \max(|\zeta^1|,|\zeta^2|)$ and let $f\colon \mathbb{C} \to F$ be the holomorphic function $f(\zeta) = (1,\zeta)$. Since $\|f(\zeta)\| = \max(1,|\zeta|)$, then $\|f(\zeta)\| = 1$ for $|\zeta| \leqslant 1$ (according to Theorem II.3.4) without f being constant on Δ.

We will now discuss conditions under which the existence of a local maximum for $\zeta \mapsto \|f(\zeta)\|$ implies that f is constant.

<u>Lemma III.1.1</u>. *Let* $f \in \mathrm{Hol}(\Delta,\mathbb{C})$ *be such that* $f(\Delta) \subset \Delta$. *Then*

$$2|z||f(0)| + (1 - |z|)|f(z) - f(0)| \leqslant 2|z|$$

for all $z \in \Delta$.

<u>Proof</u>. By the maximum principle (for scalar valued holomorphic

functions of one complex variable) $f(\Delta) \subset \Delta$ unless $f(z) = e^{i\theta}$ for some $\theta \in \mathbb{R}$, and for all $z \in \Delta$. Hence we assume $f(\Delta) \subset \Delta$. By the Schwarz-Pick lemma

$$\left| \frac{f(z) - f(0)}{1 - \overline{f(0)}f(z)} \right| \leqslant |z| \qquad (|z| < 1).$$

Since

$$|1 - \overline{f(0)}f(z)| \leqslant |1 - |f(0)|^2| + |f(0)||f(0) - f(z)| =$$

$$= 1 - |f(0)|^2 + |f(0)||f(z) - f(0)|,$$

then

$$|f(z)-f(0)| \leqslant |z||1-\overline{f(0)}f(z)| \leqslant |z|(1-|f(0)|^2) +$$

$$+ |z||f(0)||f(z)-f(0)|,$$

i.e.

$$(1 - |z||f(0)|)|f(z) - f(0)| \leqslant |z|(1 - |f(0)|^2) \leqslant$$

$$\leqslant |z|(1 + |f(0)|)(1 - |f(0)|) \leqslant 2|z|(1 - |f(0)|),$$

or

$$2|z||f(0)| + (1 - |z||f(0)|)|f(z) - f(0)| \leqslant 2|z|.$$

Since

$$1 - |z| \leqslant 1 - |z||f(0)|,$$

the last inequality proves the lemma.

Q.E.D.

<u>Lemma III.1.2</u>. *Let* B *be the open unit ball in the complex normed space* F, *and let* $f \in \mathrm{Hol}(\Delta, F)$ *be such that* $f(\Delta) \subset \overline{B}$ *(the closure of* B*). Then*

$$\|f(0) + \zeta(f(z) - f(0))\| \leqslant 1$$

for all $z \in \Delta\setminus\{0\}$ *and all* $\zeta \in \mathbb{C}$ *such that* $|\zeta| \leqslant \frac{1-|z|}{2|z|}$.

<u>Proof</u>. If $F = \mathbb{C}$ the inequality follows trivially from Lemma III.1.1. In general, suppose that

$$\|f(0) + \zeta(f(z) - f(0))\| > 1$$

for some $z \in \Delta \setminus \{0\}$ and some $|\zeta| \leqslant \frac{1-|z|}{2|z|}$. By the Hahn-Banach theorem there is a continuous linear form λ on F such that $\|\lambda\| = 1$, and

$\lambda(f(0) + \zeta(f(z) - f(0))) = \|f(0) + \zeta(f(z) - f(0))\| > 1.$

The function $\lambda \circ f$ being holomorphic on Δ, and being $\lambda \circ f(\Delta) \subset \overline{\Delta}$, this contradicts Lemma III.1.1.

Q.E.D.

The familiar notion of real extreme point is generalized by the

Definition. Let $K \subset F$. A point $x \in K$ is called a *complex extreme point* of K, if $y=0$ is the only vector in F such that $x + \zeta y \in K$ for all $\zeta \in \Delta$.

Under the hypotheses of Lemma III.1.2, if $f(z) \neq f(0)$ at some point $z \in \Delta$, the image of Δ by the (non-constant) map $\zeta \mapsto f(0) + \zeta\frac{1-|z|}{2|z|}(f(z)-f(0))$ is contained in $\overline{B}$. Thus, if $\|f(0)\| = 1$, $f(0)$ is not a complex extreme point of $\overline{B}$. That proves the first part of the following proposition. The second part is a trivial consequence of the foregoing definition.

Proposition III.1.3. *Let* $f \in \mathrm{Hol}(\Delta, F)$ *be such that* $f(\Delta) \subset \overline{B}$. *If* $f(z_0)$ *is a complex extreme point of* $\overline{B}$ *for some* $z_0 \in \Delta$, *then* f *is constant. Viceversa, if* $x \in \overline{B}$ *is not a complex extreme point of* $\overline{B}$ *then there exists a non-constant* $f \in$ $\in \mathrm{Hol}(\Delta, F)$ *such that* $f(\Delta) \subset \overline{B}$ *and* $f(0) = x$.

This proposition yields the following theorem, due to E. Thorp and R. Whitley.

Theorem III.1.4. *(Strong maximum principle). Let* D *be a domain in the complex normed space* E, *and let* $f \in \mathrm{Hol}(D, F)$

be such that $f(D) \subset \overline{B}$. *If every vector of norm one in* F *is a complex extreme point of* $\overline{B}$, *then either* $f(D) \subset B$ *or* f *is constant.*

Proof. Let $u \in D$ be such that $\|f(u)\| = 1$, and let $r > 0$ be such that $B(u,r) \subset D$. Applying the above proposition to the holomorphic function $\zeta \mapsto f(u + \zeta x)$ with $x \in B(0,r)$ we see that $f(u) = f(u+x)$. Thus f is constant on $B(u,r)$, and therefore is constant on D.

Q.E.D.

Corollary III.1.5. *If every vector of norm one in* F *is a complex extreme point of* $\overline{B}$, *for any* $f \in \mathrm{Hol}(D,F)$ *the function* $x \mapsto \|f(x)\|$ *cannot reach a relative maximum on* D *without* f *being constant.*

If x is a real extreme point of a convex set $K \subset F$, x is also a complex extreme point of K. However the converse is not true in general. Here is an example. Let Ξ be a σ-algebra of subsets of a set M, and let μ be a positive measure on M. Let B be the open unit ball of the complex Banach space $F = L^1(M,\Xi,\mu)$. We will prove the following proposition, due to E. Thorp and R. Whitley.

Proposition III.1.6. *Every point with norm one in* $L^1(M,\Xi,\mu)$ *is a complex extreme point of* $\overline{B}$.

We begin by establishing two lemmas. For any $x \in L^1(M,\Xi,\mu)$ let $S(x) = \{m \in M\colon x(m) \neq 0\}$. Note that $S(x)$ is measurable.

Lemma III.1.7. *Let* $x,y \in L^1(M,\Xi,\mu)$. *Then* $\|x+y\| = \|x\| + \|y\|$, *if, and only if, there is a positive function* h *on* $S(x) \cap$

$\cap S(y)$ *such that* $y = hx$ *almost everywhere on* $S(x) \cap S(y)$.

Proof. Let h be a function on $S(x) \cap S(y)$ such that $h(m) > 0$ for all $m \in S(x) \cap S(y)$, and $y = hx$ almost everywhere on $S(x) \cap S(y)$.

Then

$$\|x\| + \|y\| = \int_{M \setminus S(x) \cap S(y)} |x| d\mu + \int_{M \setminus S(x) \cap S(y)} |y| d\mu +$$

$$+ \int_{S(x) \cap S(y)} (1+h)|x| d\mu = \|x+y\|.$$

Conversely, if $\|x+y\| = \|x\| + \|y\|$, i.e. if

$$\int_{\mu} (|x+y| - |x| - |y|) d\mu = 0,$$

then $|x(m)+y(m)| = |x(m)| + |y(m)|$ for almost all $m \in M$.

Hence there is a positive function h on $S(x) \cap S(y)$ for which $y=hx$ almost everywhere on $S(x) \cap S(y)$.

Q.E.D.

Lemma III.1.8. *Let* $x, y \in L^1(M, \Xi, \mu)$ *be such that*

$$\|x\| = 1, \ \|x + \zeta y\| \leqslant 1 \quad \textit{for all} \quad \zeta \in \Delta.$$

Then $\|x + \zeta y\| = 1$ *for all* $\zeta \in \bar{\Delta}$, *and furthermore*

$$S(y) \subset S(x) \cup P,$$

where $P \in \Xi$ *and* $\mu(P) = 0$.

Proof. The first assertion follows from Theorem II.3.4. Furthermore

$$1 = \|x\| = \int_{S(x)} |x| d\mu \leqslant \frac{1}{2} \int_{S(x)} (|x+y| + |x-y|) d\mu \leqslant$$

$$\leqslant \frac{1}{2} \int_{M} (|x+y| + |x-y|) d\mu = \frac{1}{2} + \frac{1}{2} = 1,$$

and therefore

$$\int_{S(x)} (|x+y| + |x-y|) d\mu = 2.$$

Since

$$2 = \int_M (|x+y| + |x-y|)d\mu = \int_{S(x)} (|x+y| + |x-y|)d\mu +$$

$$+ 2\int_{M\setminus S(x)} |y|d\mu ,$$

then

$$\int_{M\setminus S(x)} |y|d\mu = 0$$

Q.E.D.

Proof of Proposition III.1.6. Let $x,y \in L^1(M,\Xi,\mu)$ be such that

$$\|x\| = 1, \quad \|x+\zeta y\| \leqslant 1 \quad \text{for all} \quad \zeta \in \overline{\Delta}.$$

By Lemma III.1.8 $\|x+\zeta y\| = 1$ for all $\zeta \in \overline{\Delta}$. Thus for any $\zeta \in \overline{\Delta}$

$$2 = \|x+\zeta y\| + \|x-\zeta y\| \geqslant \|x+\zeta y+x-\zeta y\| = 2\|x\| = 2$$

and therefore

$$\|x+\zeta y\| + \|x-\zeta y\| = \|x+\zeta y+x-\zeta y\| \quad \text{for all} \quad \zeta \in \overline{\Delta}.$$

By Lemma III.1.7 for any $\zeta \in \overline{\Delta}$ there is a positive function h_ζ on $S(x+\zeta y) \cap S(x-\zeta y)$ such that

$$x+\zeta y = h_\zeta(x-\zeta y),$$

i.e.

$$\zeta y = \frac{(h_\zeta - 1)x}{h_\zeta + 1} \quad \text{almost everywhere on} \quad S(x+\zeta y) \cap S(x-\zeta y).$$

For any $m \in S(x)$ there is a neighborhood A of 0 in Δ such that for all $\zeta \in A$, $(x+\zeta y)(m) \neq 0$, that is, $m \in S(x+\zeta y)$. For all $\zeta \in A$

$$\zeta\overline{x(m)}\,y(m) = \frac{h_\zeta(m)-1}{h_\zeta(m)+1}|x(m)|^2.$$

Since the right hand side is real for all $\zeta \in A$, then we must have $y(m) = 0$. Thus $y=0$ almost everywhere on $S(x)$. Since by Lemma III.1.8 $S(y) \subset S(x) \cup P$, with $\mu(P)=0$ then $y=0$

almost everywhere on M.

Q.E.D.

Remark 1. Recall that a normed vector space E is said to be *strictly normed* if $\|x+y\| = \|x\| + \|y\|$ implies that $y=tx$ for some $t > 0$, or else $x=0$. This definition may be compared with Lemma III.1.7. The space E is strictly normed if, and only if, the closed unit ball $\overline{B}$ of E is *rotund*, i.e. every vector of norm one is a real extreme point of $\overline{B}$. For example, a pre-Hilbert space is strictly normed.

For $1 < p < \infty$ $L^p(M,\Xi,\mu)$ is strictly normed. Thus, by Proposition III.1.6,

Every point with norm one in $L^p(M,\Xi,\mu)$ *is a complex extreme point of the closed unit ball, for* $1 \leqslant p < \infty$.

For example the closed unit ball of $L^1(0,1)$ for the Lebesgue measure on the unit interval, has no real extreme point. However, according to Proposition III.1.6, every vector $x \in L^1(0,1)$ with norm one is a complex extreme point of the closed unit ball.

Exercises. 1. Prove that the closed unit ball of the Banach space c_0 has no complex extreme point.

2. Let X be a locally compact, non-compact, Hausdorff space, and let $C_0(X)$ be the Banach algebra of all continuous complex-valued functions vanishing at infinity on X, with the uniform norm. Prove that the closed unit ball of $C_0(X)$ has no complex extreme point.

3. Let X be a compact Hausdorff space, and let $C(X)$ be the Banach algebra of all continuous complex valued functions on X

endowed with the uniform norm. Show that an element f in the closed unit ball of $C(X)$ is a complex extreme point of $\overline{B}$ if, and only if, $|f(x)| = 1$ for all $x \in X$.

We shall now discuss briefly extreme points in a Banach algebra. We begin with the following theorem due to S.Kakutani.

Theorem III.1.9. *Let* A *be a complex Banach algebra with an identity* e. *Then* e *is a real extreme point of the closed unit ball of* A.

Proof. For every $x \in A$, let T_x be the linear map of A into itself defined by $T_x(y) = xy$. The map $x \mapsto T_x$ is an isometric isomorphism of A into a closed Banach subalgebra of the Banach algebra $L(A)$ of all continuous linear endomorphisms of A. Thus it suffices to prove that the identity I is a real extreme point of the closed unit ball of $L(A)$.

Let $T \in L(A)$ be such that $\|I+tT\| \leqslant 1$ for $-1 \leqslant t \leqslant 1$. If A' is the dual space of A and $T^{\star} \in L(A')$ is the adjoint of T, then

$$\|I+tT^{\star}\| = \|I+tT\| \leqslant 1 \quad \text{for} \quad -1 \leqslant t \leqslant 1.$$

For $x' \in A'$, let

$$x'_t = (I + tT^{\star})x'.$$

Then

$$x' = \frac{1}{2}(x'_t + x'_{-t}) \quad \text{and} \quad \|x'_t\| \leqslant \|I+tT^{\star}\|\|x'\| \leqslant \|x'\|.$$

Thus, if x' is an extreme point of the closed unit ball of A' then $x'_t = x'$ for $-1 \leqslant t \leqslant 1$, i.e. $T^{\star}x' = 0$. By the Krein-Milman theorem $T^{\star}x' = 0$ for all $x' \in A'$, hence $T^{\star} = 0$, and therefore $T = 0$.

Q.E.D.

§ 2. A Schwarz lemma.

We shall now discuss a "Schwarz lemma" for bounded domains in a complex Banach space. The following result, together with Proposition III.2.2, was obtained by H. Cartan in 1932 for bounded domains in $\mathbb{C}^2$. However Cartan's proof holds, with no substantial change, for a bounded domain in a complex normed space E.

Proposition III.2.1. *Let* D *be a bounded domain in* E, *and let* $f: D \to D$ *be a holomorphic map. If, for some* $x_0 \in D$, $f(x_0) = x_0$, *and* $df(x_0) = Id$, *then* f *is the identity map.*

Proof. Assume $x_0 = 0$, and let

$$f(x) = x + P_{q_0}(x) + P_{q_{0+1}}(x) + \dots$$

be the power series expansion of f on a balanced neighborhood of 0 in D. Here $P_q \in \mathcal{P}^q(E,E)$, for $q \geqslant q_0 \geqslant 2$. Let $n \geqslant 2$ and let $f^n = f \circ \dots \circ f$ (n times). It is readily checked that

$$f^n(x) = x + nP_{q_0}(x) + \dots$$

on a neighborhood of 0. By Proposition II.3.6

$$\| nP_{q_0} \| \leqslant \frac{1}{r^{q_0}} \sup\{\| f^n(y)\| : y \in D\} ,$$

where r is the distance of 0 from $E \setminus D$. Since D is bounded, then there is a finite $k > 0$ such that

$$\| f^n(y)\| \leqslant k \qquad \text{for all } y \in D \text{ and } n=1,2,\dots .$$

Hence

$$n\| P_{q_0} \| \leqslant \frac{k}{r^{q_0}} \qquad \text{for } n=1,2,\dots ,$$

and therefore $P_{q_0} = 0$.

Q.E.D.

Let D be a domain in E, and D' a domain in F. We denote by $\mathrm{Hol}(D,D')$ the set of $f \in \mathrm{Hol}(D,F)$ such that $f(D) \subset \subset D'$.

An *automorphism* of D is a homeomorphism f of D onto D such that $f \in \mathrm{Hol}(D,D)$, $f^{-1} \in \mathrm{Hol}(D,D)$. By $\mathrm{Aut}(D)$ we denote the group of all automorphism of D. A domain D is said to be *circular* if, for all $x \in D$, $e^{i\theta}x \in D$ for all $\theta \in \mathbb{R}$, i.e. $\ell_\theta : x \mapsto e^{i\theta}x$ is an automorphism of D.

Proposition III.2.2. *Let* D *be a bounded circular domain containing* 0, *and let* $f \in \mathrm{Aut}(D)$ *be such that* $f(0) = 0$. *Then* f *is (the restriction to* D *of) a linear isomorphism of* E.

Proof. Let

$$f(x) = P_1(x) + P_2(x) + \dots$$

be the power series expansion of f in a neighborhood of 0. Hence P_1 is an isomorphism of E onto E. Let $g \in \mathrm{Aut}(D)$ be defined by

$$g = f^{-1} \circ \ell_{-\theta} \circ f \circ \ell_\theta .$$

Then $g(0) = 0$, and the linear part $dg(0)$ of g at 0 is expressed by

$$dg(0) = (P_1)^{-1} \circ \ell_{-\theta} \circ P_1 \circ \ell_\theta = \mathrm{Id}.$$

By Proposition III.2.1 g is the identity, i.e.

$$f \circ \ell_\theta = \ell_\theta \circ f.$$

Thus there is a neighborhood A of 0 such that

$$e^{i\theta}\{P_1(x) + P_2(x) + \dots\} = e^{i\theta}P_1(x) + e^{2i\theta}P_2(x) + \dots$$

for all $\theta \in \mathbb{R}$, and for all $x \in A$. Hence, $P_q = 0$ for $q \geqslant 2$.

Q.E.D.

Theorem III.2.3. *Let* E *and* F *be complex normed spaces, let*

B_E *and* B_F *be their open unit balls and let* $f \in Hol(B_E, F)$ *be such that* $f(0)=0$ *and* $f(B_E) \subset \overline{B_F}$. *Then*

$$\text{(III.2.1)} \qquad \|f(x)\| \leq \|x\| \qquad \textit{for all} \quad x \in B_E.$$

If $\|f(x_0)\| = \|x_0\|$ *at some point* $x_0 \in B_E \setminus \{0\}$, *then*

$$\text{(III.2.2)} \qquad \|f(\zeta x_0)\| = \|\zeta x_0\| \qquad \textit{for all} \quad |\zeta| < \frac{1}{\|x_0\|}.$$

If moreover the set $\{\frac{e^{i\theta}}{\|f(x_0)\|} f(x_0) : \vartheta \in \mathbb{R}\}$ *contains some complex extreme point of* $\overline{B_F}$, *then*

$$f(\zeta x_0) = \zeta f(x_0) \qquad \textit{for all} \quad |\zeta| < \frac{1}{\|x_0\|}.$$

<u>Proof</u>. The proof imitates that of the classical Schwarz lemma. Let $x \in B_E \setminus \{0\}$. The function $\zeta \mapsto \frac{1}{\zeta} f(\zeta x)$ is a holomorphic map of $\{\zeta \in \mathbb{C}: |\zeta| < \frac{1}{\|x\|}\}$. For $0 < r < \frac{1}{\|x\|}$ and $|\zeta| = r$

$$\text{(III.2.3)} \qquad \|\frac{1}{\zeta} f(\zeta x)\| \leq \frac{1}{r},$$

and, by the maximum principle, (III.2.3) holds for $|\zeta| \leq r$, i.e.

$$\|f(\zeta x)\| \leq \frac{|\zeta|}{r} \quad \text{for all} \quad |\zeta| \leq r, \quad 0 < r < \frac{1}{\|x\|}.$$

Letting $r \nearrow \frac{1}{\|x\|}$ we obtain $\|f(\zeta x)\| \leq |\zeta| \|x\|$ for all $|\zeta| < \frac{1}{\|x\|}$, and finally, for $\zeta = 1$, inequality (III.2.1).

If $\|f(x_0)\| = \|x_0\|$, taking $\zeta = 1$ and $x = x_0$, Theorem II.3.4 yields

$$\|\frac{1}{\zeta} f(\zeta x_0)\| = \|x_0\| \quad \text{for all} \quad |\zeta| \leq r, \qquad 0 < r < \frac{1}{\|x_0\|}.$$

Letting as before $r \nearrow \frac{1}{\|x_0\|}$ (III.2.2) follows.

If the circle $\{\frac{e^{i\theta}}{\|f(x_0)\|} f(x_0) : \theta \in \mathbb{R})$ contains a complex extreme point of $\overline{B_F}$ (and therefore all the points of the circle are complex extreme points of $\overline{B_F}$), by Proposition III.1.3 the function

$\zeta \mapsto \frac{1}{\zeta\|x_0\|} f(\zeta x_0)$ is constant for $(0<|\zeta|<\frac{1}{\|x_0\|})$, i.e. there exists a vector $y_0 \in F$, such that

$$\frac{1}{\zeta\|x_0\|} f(\zeta x_0) = y_0$$

or, equivalently,

$$f(\zeta x_0) = \zeta\|x_0\| y_0 \qquad \text{for all} \quad |\zeta| < \frac{1}{\|x_0\|} .$$

Since $\|x_0\| y_0 = f(x_0)$ the conclusion follows.

Q.E.D.

Let $h \in \mathrm{Hol}(\Delta,\mathbb{C})$ be such that $h(\Delta) \subset \bar{\Delta}$. If $h(\Delta) \not\subset \Delta$, then h is constant. Suppose that h is not constant. The holomorphic function $g: \zeta \mapsto \frac{h(\zeta)-h(0)}{1-\overline{h(0)}h(\zeta)}$ maps Δ into $\bar{\Delta}$ and 0 into 0. It follows from the power series expansion of g at 0 that $|g'(0)| \leqslant 1$, being $|g'(0)| = 1$, if and only if, $g(\zeta) = e^{i\theta}\zeta$, i.e. if, and only if, $h \in \mathrm{Aut}(\Delta)$.

Being

$$g'(0) = \frac{h'(0)}{1-|h(0)|^2} ,$$

then

$$|h'(0)| + |h(0)|^2 \leqslant 1, \tag{III.2.4}$$

equality holding if, and only if, h is a Moebius transformation.

We prove now the following theorem, due to L.A. Harris[1].

<u>Theorem III.2.4</u>. *Let* $f \in \mathrm{Hol}(B_E,F)$ *be such that* $f(B_E) \subset B_F$. *If* $df(0)$ *is an isometry of* E *onto* F, *then* f *is linear, i.e.* f *is the restriction of* $df(0)$ *to* B .

<u>Proof.</u> Replacing f by $df(0)^{-1} \circ f$, we may assume $f \in$ $\in \mathrm{Hol}(B_E,E)$, $f(B_E) \subset \overline{B_E}$, $df(0) = \mathrm{Id}$. Theorem III.2.4 will then follow from Proposition III.2.1 once we show that $f(0)=0$.

Suppose that $x_0 = f(0) \neq 0$. By the Hahn-Banach theorem, there exists a continuous linear form λ on E such that $\|\lambda\| = 1$, and $\lambda(x_0) = \|x_0\|$. Let $h: \Delta \to \mathbb{C}$ be the holomorphic function defined by

$$h(\zeta) = \lambda(f(\frac{\zeta}{\|x_0\|} x_0)) \quad (\zeta \in \Delta).$$

Being

$$|h(\zeta)| \leqslant \|\lambda\| \|f(\frac{\zeta}{\|x_0\|} x_0)\| \leqslant 1$$

then $h(\Delta) \subset \overline{\Delta}$. Since $h(0) = \lambda(x_0) = \|x_0\|$, $h'(0) = \lambda(\frac{1}{\|x_0\|}x_0) = 1$ (hence h is non-constant), (III.2.4) yields $\|x_0\|^2 \leqslant 1-1 = 0$, contradicting the assumption $x_0 \neq 0$.

Q.E.D.

Notes.

The notion of complex extreme point was introduced by E. Thorp and R. Whitley [1], who proved also the strong maximum principle (Theorem III.1.4). Our proof is a simplification of their original proof and is due to L. A. Harris [1]. Also Theorem III.2.4 is due to L. A. Harris [1]. Our proof relies on an extension of the uniqueness theorem (Proposition III.2.1) established by H. Cartan [H.Cartan, 1, S.Bochner-W.T.Martin,1, p.13] in the finite-dimensional case. As it was noticed by L. A. Harris [1, p.1017], Cartan's proof extends with no substantial modification to the infinite dimensional case. Proposition III.2.2 is also due to H. Cartan [1].

The description of the complex-extreme points of $L^1(M,\Xi,\mu)$ (Proposition III.1.6) is due to Thorp and Whitley [1]. For the description of the complex extreme points of the closed unit ball of $L^p(M,\Xi,\mu)$, cf. e.g. Dunford and Schwartz [1].

Let A be a $C^\star$ algebra and let $\overline{B}$ be its closed unit ball. Then $\overline{B}$ has real extreme points if, and only if, A has an identity [Sakai 1, p.10-11]. If A has an identity, e, the set of real extreme points of $\overline{B}$ consists of all partial isometries x of A in A such that

$$(e - xx^\star)A(e - x^\star x) = \{0\}.$$

This result is due to R. V. Kadison [1; cf. [S.Sakai, 1, p.12]. The set of complex extreme points of $\overline{B}$ coincides with the set of real extreme points. This fact was established by C. A. Ackermann and B. Russo [1], where also the case of a von Neumann algebra and of its predual was considered. As a consequence, the strong maximum principle fails in every $C^\star$-algebra of dimension larger than one. This fact had already been noticed in [Brown-Douglas, 1] in a particular case.

Special types of real and complex extreme points have been investigated by H.F. Bohnenblust and S. Karlin [1], G. Lumer [1], R. McGuigan [1] and L.A. Harris [3].

For extensions to holomorphic functions on domains in locally convex spaces cf. E. Vesentini [4].

CHAPTER IV

INVARIANT PSEUDODISTANCES

§ 1. The Kobayashi and Carathéodory pseudodistances.

Let D be a domain in a normed vector space E over $\mathbb{C}$. Let x, y be two points of D. An *analytic chain* joining x and y in D consists of 2ν points $\zeta_1', \zeta_1'', \ldots, \zeta_\nu', \zeta_\nu''$ in Δ and of ν functions $f_j \in \mathrm{Hol}(\Delta, D)$, such that

$$f_1(\zeta_1') = x, \ldots, f_j(\zeta_j'') = f_{j+1}(\zeta_{j+1}') \quad \text{for} \quad j=1,\ldots,\nu-1, \; f_\nu(\zeta_\nu'') = y.$$

Since D is connected, given x and y, an analytic chain joining x and y in D always exists, provided that ν is sufficiently large. Let

$$\text{(IV.1.1)} \qquad k_D(x,y) = \inf\{\omega(\zeta_1', \zeta_1'') + \omega(\zeta_2', \zeta_2'') + \ldots + \omega(\zeta_\nu', \zeta_\nu'')\},$$

where ω is the Poincaré distance in Δ and the infimum is taken over all choices of analytic chains joining x and y in D.

Clearly

$$k_D(x,y) = k_D(y,x) \geqslant 0, \quad k_D(x,x) = 0, \quad k_D(x,y) \leqslant k_D(x,z) + k_D(z,y)$$

for all choices of x, y and z in D. In other words the function $k_D\colon D \times D \to \overline{\mathbb{R}}_+$ is a pseudo-distance on D. It is called the *Kobayashi pseudo-distance*, or the *Kobayashi distance* if $k_D(x,y) > 0$ whenever $x \neq y$.

Remark. In the above construction there is no restriction in assuming

$$\zeta_1' = 0, \quad \zeta_1'' = \zeta_2', \dots, \zeta_j'' = \zeta_{j+1}', \dots, \zeta_{\nu-1}'' = \zeta_\nu' .$$

Indeed let g_1 be a Möbius transformation such that $g_1(0) = \zeta_1'$, and let $\eta_1 = g_1^{-1}(\zeta'')$. Recurrently let g_{j+1} be the Möbius transformation such that $g_{j+1}(\eta_j) = \zeta_{j+1}$ and let $\eta_{j+1} = g_{j+1}^{-1}(\zeta_{j+1}'')$ for $j=1,\dots,\nu-1$.

Consider the holomorphic maps $h_j = f_j \circ g_j : \Delta \to D$.

We have

$$h_1(0) = f_1(g_1(0)) = f_1(\zeta_1') = x,$$

$$h_2(\eta_1) = f_2(g_2(\eta_1)) = f_2(\zeta_2') = f_1(\zeta_1'') = f_1(g_1(\eta_1)) = h_1(\eta_1),$$

$$\dots$$

$$h_{j+1}(\eta_j) = f_{j+1}(g_{j+1}(\eta_j)) = f_{j+1}(\zeta_{j+1}') = f_j(\zeta_j'') = f_j(g_j(\eta_j)) = h_j(\eta_j),$$

$$\dots$$

$$h_\nu(\eta_\nu) = f_\nu(g_\nu(\eta_\nu)) = f_\nu(\zeta_\nu'') = y.$$

Furthermore

$$\omega(0,\eta_1) + \omega(\eta_1,\eta_2) + \dots + \omega(\eta_j,\eta_{j+1}) + \dots + \omega(\eta_{\nu-1},\eta_\nu) =$$
$$= \omega(g_1(0),g_1(\eta_1)) + \omega(g_2(\eta_1),g_2(\eta_2) + \dots + \omega(g_{j+1}(\eta_j),g_{j+1}(\eta_{j+1})) + \dots$$
$$\dots + \omega(g_\nu(\eta_{\nu-1}),g_\nu(\eta_\nu)) =$$
$$= \omega(\zeta_1',\zeta_1'') + \omega(\zeta_2',\zeta_2'') + \dots + \omega(\zeta_{j+1}',\zeta_{j+1}'') + \dots + \omega(\zeta_\nu',\zeta_\nu'').$$

Hence we may state the following

Lemma IV.1.1. *Given* x *and* y *in* D, *there exist: a positive integer* ν; ν *points* $\zeta_1,\dots,\zeta_\nu$ *in* Δ *and* ν *holomorphic maps* $f_1,\dots,f_\nu$ *of* Δ *into* D *such that*

$$f_1(0) = x, \quad f_j(\zeta_j) = f_{j+1}(\zeta_j) \quad \textit{for} \quad j=1,\dots,\nu-1, \quad f_\nu(\zeta_\nu) = y.$$

Furthermore

$$k_D(x,y) = \inf\{\omega(0;\zeta_1) + \omega(\zeta_1,\zeta_2) + \dots + \omega(\zeta_{\nu-1},\zeta_\nu)\},$$

where the infimum is taken over all choices of ν,
$\zeta_1, \dots, \zeta_\nu$, $f_1, \dots, f_\nu$.

The definition of k_D implies

Proposition IV.1.2. *Let* D_1 *be a domain in a complex normed space* E_1, *and let* $F: D \to D_1$ *be a holomorphic map. Then for all* x, y *in* D

$$k_{D_1}(F(x), F(y)) \leqslant k_D(x,y).$$

Corollary IV.1.3. *If* $F: D \to D_1$ *is a bi-holomorphic map, then*

$$k_{D_1}(F(x), F(y)) = k_D(x,y), \quad \textit{for all } x, y \textit{ in } D.$$

In particular, every automorphism of D is an isometry for k_D.

Corollary IV.1.4. *Let* D *and* D' *be two domains in* E, *such that* $D \subset D'$. *Then*

$$k_{D'}(x,y) \leqslant k_D(x,y) \qquad \textit{for all } x,y \textit{ in } D.$$

We compute now the Kobayashi distance k_Δ on Δ. Taking in the definition $\nu=1$, $f(\zeta)=\zeta$, we have

$$k_\Delta(x,y) \leqslant \omega(x,y) \qquad (x,y \in \Delta).$$

On the other hand, by the Schwarz-Pick lemma, and - setting $f_j(\zeta_j'') = f_{j+1}(\zeta_{j+1}')$ - by the triangle inequality,

$$\omega(\zeta_1',\zeta_1'') + \omega(\zeta_2',\zeta_2'') + \dots + \omega(\zeta_\nu',\zeta_\nu'') \geqslant \omega(f_1(\zeta_1'),f_1(\zeta_1'')) +$$
$$+ \omega(f_2(\zeta_2'),f_2(\zeta_2'')) + \dots + \omega(f_\nu(\zeta_\nu'),f_\nu(\zeta_\nu'')) \geqslant$$
$$\geqslant \omega(f_1(\zeta_1'),f_\nu(\zeta_\nu'')) = \omega(x,y),$$

for every choice of the analytic chain joining x and y in Δ. Thus $k_\Delta(x,y) \geqslant \omega(x,y)$, and in conclusion

$$k_\Delta(x,y) = \omega(x,y) \qquad \text{for all } x,y \text{ in } \Delta. \tag{IV.1.2}$$

According to the following proposition, the Kobayashi pseudo-distance is the "largest" pseudo-distance on D for which every holomorphic map $f: \Delta \to D$ is a contraction.

<u>Proposition IV.1.5</u>. *Let* d *be a pseudo-distance on* D *such that*

$$d(f(\zeta_1), f(\zeta_2)) \leqslant \omega(\zeta_1, \zeta_2)$$

for all holomorphic maps $f: \Delta \to D$ *and all points* ζ_1, ζ_2 *in* Δ. *Then*

$$d(x,y) \leqslant k_D(x,y)$$

for all x, y *in* D.

<u>Proof</u>. With the same notations as in the definition of k_D, the triangle inequality yields

$$d(x,y) \leqslant \sum_{j=1}^{\nu} d(f_j(\zeta_j'), f_j(\zeta_j'')) \leqslant \sum_{j=1}^{\nu} \omega(\zeta_j', \zeta_j'')$$

for every choice of $\zeta_1', \zeta_1'', \dots, \zeta_\nu', \zeta_\nu'', f_1, \dots, f_\nu$. Thus

$$d(x,y) \leqslant k_D(x,y).$$

Q.E.D.

By Proposition IV.1.2 and by (IV.1.2),

$$\omega(f(x), f(y)) \leqslant k_D(x,y)$$

for all x, y in D and all $f \in \mathrm{Hol}(D, \Delta)$. Thus, letting

$$c_D(x,y) = \sup\{\omega(f(x), f(y)) : f \in \mathrm{Hol}(D,\Delta)\}, \tag{IV.1.3}$$

then

$$c_D(x,y) \leqslant k_D(x,y). \tag{IV.1.4}$$

Clearly

$$c_D(y,x) = c_D(x,y) \geqslant 0, \quad c_D(x,y) = 0, \quad c_D(x,y) \leqslant c_D(x,z) + c_D(z,y)$$

for all x, y, z in D. Thus $c_D : D \times D \to \mathbb{R}_+$ is a pseudo-distance. It is called the *Carathéodory pseudo-distance* on D or the *Carathéodory distance* if $c_D(x,y) > 0$ whenever $x \neq y$.

The definition (IV.1.3) implies immediately

Proposition IV.1.6. *Proposition IV.1.2 and Corollaries IV.1.3 and IV.1.4 hold with* c_D *and* c_{D_1} *in place of* k_D *and* k_{D_1}.

By definition $c_\Delta(x,y) \geqslant \omega(x,y)$. Thus, by (IV.1.4) and (IV.1.2) we have

$$c_\Delta(x,y) = \omega(x,y) \qquad \forall\ x,y \text{ in } \Delta. \tag{IV.1.5}$$

The Carathéodory pseudo-distance is the "smallest" pseudo-distance in D for which every holomorphic map $f: D \to \Delta$ is a contraction. In fact the following proposition is an obvious consequence of the definition (IV.1.3).

Proposition IV.1.7. *Let* $d: D \times D \to \mathbb{R}_+$ *be a pseudo-distance on* D *such that*

$$d(x,y) \geqslant \omega(f(x),f(y))$$

for all x,y *in* D *and every* $f \in \mathrm{Hol}(D,\Delta)$. *Then*

$$c_D(x,y) \leqslant d(x,y).$$

Theorem IV.1.8. *Let* p *be a continuous semi-norm on* E, *and let*

$$B_p = \{x \in E: p(x) < 1\}.$$

Then

$$c_{B_p}(0,x) = k_{B_p}(0,x) = \omega(0,p(x))$$

for all $x \in B_p$.

Proof. Let $x \in B_p$ with $p(x) > 0$. The holomorphic function $\zeta \to \frac{\zeta}{p(x)}x$ maps Δ into B_p, 0 into 0, and $p(x)$ into x. Thus

$$c_{B_p}(0,x) \leqslant k_{B_p}(0,x) \leqslant \omega(0,p(x)).$$

On the other hand there exists a continuous linear form λ in E such that $\lambda(x) = p(x)$ and $|\lambda(y)| \leqslant p(y)$ for all $y \in E$. Thus $\lambda \in \mathrm{Hol}(D,\Delta)$, and

$$\omega(0,p(x)) = \omega(0,\lambda(x)) \leqslant c_{B_p}(0,x).$$

Let $x \neq 0$, but $p(x) = 0$. Choose any $t > 1$. The holomorphic function $f : \zeta \to t\zeta x$ maps Δ into B_p, and moreover $f_t(0) = 0$, $f_t(\frac{1}{t}) = x$. Hence

$$c_{B_p}(0,x) \leqslant k_{B_p}(0,x) \leqslant \omega(0,\tfrac{1}{t}).$$

Letting $t \to \infty$, we get

$$c_{B_p}(0,x) = k_{B_p}(0,x) = 0.$$

The proof of the lemma is complete.

Q.E.D.

<u>Corollary IV.1.9</u>. *Let* $B_p(x_0,r)$ *be the open ball with center* x_0 *and radius* r, *with respect to a continuous semi-norm* p *of* E. *Then*

$$c_{B_p(x_0,r)}(x_0,x) = k_{B_p(x_0,r)}(x_0,x) = \omega(0, \frac{p(x-x_0)}{r}) \tag{IV.1.6}$$

for all $x \in B_p(x_0,r)$.

<u>Examples</u>. 1. Let $E = \mathbb{C}^2$ and $\|(\zeta^1,\zeta^2)\| = \max(|\zeta^1|,|\zeta^2|)$. then $B((0,0),1) = \Delta \times \Delta$, and

$$c_{\Delta\times\Delta}((0,0),(\zeta^1,\zeta^2)) = k_{\Delta\times\Delta}((0,0),(\zeta^1,\zeta^2)) = \max\{\omega(0,\zeta^1),\omega(0,\zeta^2)\}.$$

Since $\mathrm{Aut}\Delta \times \mathrm{Aut}\Delta$ acts transitively on $\Delta \times \Delta$, then we have

$$c_{\Delta\times\Delta}((\zeta^1,\zeta^2),(\zeta'^1,\zeta'^2)) = k_{\Delta\times\Delta}((\zeta^1,\zeta^2),(\zeta'^1,\zeta'^2)) = \max\{\omega(\zeta^1,\zeta'^1),\omega(\zeta^2,\zeta'^2)\}.$$

2. Choosing the semi-norm $p = 0$, Corollary IV.1.9 yields

$$c_E(x,y) = k_E(x,y) = 0$$

for all x, y in E.

For $0 \leq t < 1$, $\omega(0,t) = \frac{1}{2}\log\frac{1+t}{1-t}$ is expressed by

$$\omega(0,t) = \frac{1}{2}\log\frac{1+t}{1-t} = \sum_{n=0}^{+\infty} \frac{t^{2n+1}}{2n+1}. \tag{IV.1.7}$$

Thus $t \mapsto \omega(0,t)$ is a convex increasing function, vanishing at $t = 0$. For $0 < s < 1$ there is a positive constant k such that

$$\omega(0,t) \leq kt \qquad \text{for } 0 \leq t \leq s. \tag{IV.1.8}$$

Lemma IV.1.10. *For every* $x_0 \in D$ *and for every* $r > 0$ *such that* $\overline{B(x_0,r)} \subset D$ *there exists a constant* $k > 0$ *such that*

$$k_D(x_0,x) \leq k\|x - x_0\|$$

for all $0 < s < r$ *and all* $x \in \overline{B(x_0,s)}$.

Proof. Being

$$\omega\left(0, \frac{\|x-x_0\|}{r}\right) = \frac{1}{2}\log\frac{1 + \frac{\|x-x_0\|}{r}}{1 - \frac{\|x-x_0\|}{r}}$$

the assertion follows from the inequality

$$k_D(x_0,x) \leq k_{B(x_0,r)}(x_0,x),$$

from Corollary IV.1.9 and from (IV.18).

Q.E.D.

Lemma IV.1.10 and (IV.1.8) imply that for every $x_0 \in D$ the functions $x \mapsto k_D(x_0,x)$, $x \mapsto c_D(x_0,x)$ are continuous. Since for $x_0, y_0, x, y \in D$

$$|k_D(x_0,y_0) - k_D(x,y)| \leq k_D(x_0,x) + k_D(y_0,y),$$

$$|c_D(x_0,y_0) - c_D(x,y)| \leqslant c_D(x_0,x) + c_D(y_0,y),$$

we obtain

Theorem IV.1.11. *The functions* $c_D\colon D\times D \to \mathbb{R}_+$, $k_D\colon D\times D\to\mathbb{R}_+$ *are continuous.*

In the next section we shall examine the topologies defined by c_D and k_D on bounded domains. Now we prove a lemma which will be useful in chapter V.

For any subset $A \subset D$ and any $r > 0$, we set

$$B_k(A,r) = \{x \in D\colon k_D(x,y) < r \text{ for some } y \in A\}.$$

Lemma IV.1.12. *Let* D *be a domain in a complex normed space* E. *For all* $x \in D$ *and all* $r > 0$, $r' > 0$,

$$B_k(B_k(x,r),r') = B_k(x,r+r').$$

Proof. By the triangle inequality

$$B_k(B_k(x,r),r') \subset B_k(x,r+r').$$

Thus we need only prove the opposite inclusion. In other words we have to show that for every $y \in B_k(x,r+r')$ there is some $z \in B_k(x,r)$ such that $k_D(y,z) < r'$. Choose $\varepsilon > 0$ such that $k_D(x,y) \leqslant r+r'-\varepsilon$ and $\varepsilon \leqslant r'$. With the same notations as in the definition of k_D, let $(\nu,\zeta_1',\ldots,f_\nu)$ be an analytic chain in D such that $f_1(\zeta_1') = x$, $f_\nu(\zeta_\nu'') = y$, and

$$\omega(\zeta_1',\zeta_1'') + \ldots + \omega(\zeta_\nu',\zeta_\nu'') < r + r' - \frac{\varepsilon}{2}\,.$$

Being

$$\omega(\zeta_1',\zeta_1'') + \ldots + \omega(\zeta_\nu',\zeta_\nu'') \triangleq r + r' - \varepsilon = r - \frac{\varepsilon}{4} + r' - \frac{3\varepsilon}{4}\,,$$

and since $r' > \frac{3}{4}\varepsilon$, then

$$\omega(\zeta_1',\zeta_1'') + \ldots + \omega(\zeta_\mu',\zeta_\mu'') \geqslant r - \frac{\varepsilon}{4}$$

for some $1 \leqslant \mu \leqslant \nu$. Choose the smallest μ for which the latter inequality holds.

Let ℓ be the geodesic arc for the Poincaré metric on Δ

joining ζ'_μ and ζ''_μ, let τ_μ be the unique point on ℓ such that

$$\text{(IV.1.9)} \qquad \omega(\zeta'_1,\zeta''_1)+\dots+\omega(\zeta'_{\mu-1},\zeta''_{\mu-1})+\omega(\zeta'_\mu,\tau_\mu) = r-\frac{\varepsilon}{4} < r,$$

and let $z = f_\mu(\tau_\mu)$. Then

$$k_D(x,z) \leq \omega(\zeta'_1,\zeta''_1)+\dots+\omega(\zeta'_{\mu-1},\zeta''_{\mu-1})+\omega(\zeta'_\mu,\tau_\mu) = r-\frac{\varepsilon}{4}<r,$$

i.e. $z \in B_k(x,r)$, while, by (IV.1.9),

$$k_D(y,z) \leq \omega(\tau_\mu,\zeta''_\mu)+\omega(\zeta'_{\mu+1},\zeta''_{\mu+1})+\dots+\omega(\zeta'_\nu,\zeta''_\nu) =$$

$$= \omega(\zeta'_\mu,\zeta''_\mu)+\omega(\zeta'_{\mu+1},\zeta''_{\mu+1})+\dots+\omega(\zeta'_\nu,\zeta''_\nu)-\omega(\zeta'_\mu,\tau_\mu) =$$

$$= \sum_{j=1}^{\nu}\omega(\zeta'_j,\zeta''_j) - \sum_{j=1}^{\mu-1}\omega(\zeta'_j,\zeta''_j) - \omega(\zeta'_\mu,\tau_\mu) < r+r'-\frac{\varepsilon}{2}-(r-\frac{\varepsilon}{4}) =$$

$$= r'-\frac{\varepsilon}{4} < r'.$$

Q.E.D.

Remark. A result similar to Lemma IV.1.12 does not hold in general for the Carathéodory pseudo-distance. In fact, consider in $\mathbb{C}^2$ the domain

$$D = \Delta\times\Delta \setminus \{z=(z_1,z_2)\,|\in\mathbb{C}^2 : |z_1|<a,\ |z_2|<1-a\},$$

where $0<a<\frac{1}{2}$.

By Hartogs' theorem every holomorphic function $f: D\to\mathbb{C}$ extends to a holomorphic mapping $\tilde{f}: \Delta\times\Delta\to\mathbb{C}$. This means that the Carathéodory distance c_D coincides with the restriction to D of the Carathéodory distance of the bidisc. Thus we have (Example 1):

$$c_D(x,y) = \max\{\omega(x_1,y_1),\ \omega(x_2,y_2)\} \quad (x=(x_1,x_2),\ y=(y_1,y_2)\in D).$$

Let now $x = (2a,0)\in D$.

First of all,

$$c_D(x,-x) = \log\frac{1+2a}{1-2a}.$$

Let $x\in D$ and let $B_c(x,r) = \{y\in D: c_D(x,y)<r\}$. If

ε is sufficiently small, there exists a positive constant b such that

$$B_c(x, \frac{1}{2}\log\frac{1+2a}{1-2a}+\varepsilon) \subset \{z = (z_1, z_2) \in D : \operatorname{Re} z > b\}.$$

$$B_c(x, \frac{1}{2}\log\frac{1+2a}{1-2a}+\varepsilon) \cap B_c(-x, \frac{1}{2}\log\frac{1+2a}{1-2a}+\varepsilon) = \emptyset;$$

on the other hand $-x \in B_c(x, \log\frac{1+2a}{1-2a}+2\varepsilon)$.

We prove now

Theorem IV.1.13. *For any* $x_0 \in D$, *the function* $x \mapsto \log c_D(x_0, x)$ *is a continuous plurisubharmonic function on* D.

We begin by considering the Poincaré distance.

Lemma IV.1.14. *For any* $\zeta_0 \in \Delta$, *the function* $\zeta \mapsto \log \omega(\zeta_0, \zeta)$ *is subharmonic on* Δ.

Proof. (C. Berenstein). The group $\operatorname{Aut}\Delta$ being transitive on Δ, there is no restriction in choosing $\zeta_0 = 0$. Let $\phi(t) = \log\log\frac{1+t}{1-t}$ $(0 \leqslant t < 1)$. Then for $0 < |\zeta| = t < 1$,

$$4\frac{\partial^2}{\partial\zeta\partial\bar\zeta}\log\omega(0,\zeta) = \phi''(t) + \frac{1}{t}\phi'(t) =$$

$$= \frac{2}{(1-t^2)^2}\left(\log\frac{1+t}{1-t}\right)^{-2}\left(\frac{1+t^2}{t}\log\frac{1+t}{1-t} - 2\right).$$

Since

$$\frac{1+t^2}{t}\log\frac{1+t}{1-t} = 2(1+t^2)\sum_{n=0}^{+\infty}\frac{t^{2n}}{2n+1},$$

then

$$\frac{\partial^2}{\partial\zeta\partial\bar\zeta}\log\omega(0,\zeta) > 0$$

for all $\zeta \in \Delta\setminus\{0\}$. Being $\omega(0,0) = 0$, and $\zeta \mapsto \omega(0,\zeta)$ being positive and continuous on Δ, then $\zeta \mapsto \log\omega(0,\zeta)$ is subharmonic on Δ.

Q.E.D.

Let $x_0 \in D$. For any $f \in \mathrm{Hol}(D,\Delta)$, the function $x \mapsto \log \omega(f(x_0),f(x))$ is a continuous plurisubharmonic function on D. Since the function $x \mapsto \log c_D(x_0,x)$ is continuous, Lemma II.6.1 yields the proof of Theorem IV.1.13.

§ 2. Hyperbolic domains.

Let U be a proper open subset of E.

A closed subset $K \subset U$ is said to be *completely interior* to U - in symbols, $K \subset\subset U$ - if $d(K,E\setminus U) > 0$. Here

$$d(K,E\setminus U) = \inf\{\|x-y\|: x \in K,\ y \in E\setminus U\},$$

is the norm-distance of K from $E\setminus U$.

Remarks. 1. If $K \subset U$ is compact, then $K \subset\subset U$.

2. If $\dim_{\mathbb{C}} E < \infty$, and if U is bounded, then K is completely interior to U (if, and only) if K is compact. That is not always the case if U is not bounded. For example the strip $\{\zeta \in \mathbb{C}: 1 \leqslant \mathrm{Im}\zeta \leqslant 2\}$ is completely interior to the upper half plane $\{\zeta \in \mathbb{C}: \mathrm{Im}\zeta > 0\}$.

3. Let $x_0 \in U$, and let $r > 0$ be such that the open ball

$$B(x_0,r) = \{x \in E: \|x-x_0\| < r\} \subset U.$$

Then $\overline{B(x_0,r)} \subset\subset U$ if, and only if, there is some $R > r$, such that $B(x_0,R) \subset U$.

Suppose now that D is bounded in the normed complex vector space E. For any $x_0 \in D$ there is a positive R such that $D \subset B(x_0,R)$. For any $x \in D$ we have, by Corollary IV.1.9 and by (IV.1.7)

$$c_D(x_0,x) \geqslant c_{B(x_0,R)}(x_0,x) = \omega\left(0,\frac{\|x-x_0\|}{R}\right) \geqslant \frac{\|x-x_0\|}{R},$$

and therefore

Lemma IV.2.1. *If* D *is a bounded domain in* E, *for every* $x_0 \in D$ *there is a positive constant* k_1 *such that*

$$k_1 \|x-x_0\| \leqslant c_D(x_0, x) \qquad \textit{for all} \quad x \in D.$$

Lemmas IV.1.10 and IV.2.1 yield

Theorem IV.2.2. *Let* D *be a bounded domain in a complex Banach space* E. *The pseudo-distancex* c_D *and* k_D *are equivalent to the norm-distance on any closed ball (for the norm) completely interior to* E. *In particular, both* k_D *and* c_D *are distances.*

Definition. If k_D is a distance and if the topology defined by k_D is equivalent to the relative topology of D in E, the domain D is said to be *hyperbolic*. By Theorem IV.2.2 any bounded domain in E is hyperbolic. If $\dim_{\mathbb{C}} E < \infty$, the fact that k_D is a distance suffices to ensure that D is hyperbolic:

Proposition IV.2.3. *If* $\dim E < \infty$, *and if* k_D *is a distance on the domain* $D \subset E$, *then* k_D *defines in* D *the relative topology*, i.e., D *is hyperbolic.*

Proof. We have to show that for every $x \in D$ and every neighborhood U of x in D (for the standard vector topology of E) there is a ball

$$B_k(x,r) = \{y \in E: k_D(x,y) < r\}$$

with positive radius r, such that $B_k(x,r) \subset U$. If that is not the case there is a sequence (x_n) in D such that $x_n \notin$ $\notin U$, and $k_D(x,x_n) < \frac{1}{n}$. Since E is finite dimensional, we can choose U relatively compact in D.

With the notations introduced at the beginning of this section, for every $n=1,2,\ldots$ there exists a positive integer

$\nu(n)$ and an analytic chain $(\nu(n), \zeta_1'^n, \dots, \zeta''^n_{\nu(n)}, f_1^n, \dots, f^n_{\nu(n)}$ joining x and x_n in D, such that

$$\omega(\zeta_1'^n, \zeta_1''^n) + \omega(\zeta_2'^n, \zeta_2''^n) + \dots + \omega(\zeta'^n_{\nu(n)}, \zeta''^n_{\nu(n)}) < \frac{1}{n} .$$

Let ℓ_j^n be the geodesic arc for the Poincaré metric on Δ, joining $\zeta_j'^n$, $\zeta_j''^n$. The maps f_j^n $(j=1, \dots, \nu(n))$ define then an arc ℓ^n in D joining x and x_n. For any $y \in \ell^n$, $k_D(x,y) < \frac{1}{n}$.

Let ∂U be the boundary of U. Since $x \in U$, $x_n \notin U$, then $\ell^n \cap \partial U$ is non-empty.

Let $y_n \in \ell^n \cap \partial U$. Then

(IV.2.1) $$\lim_{n \to \infty} k_D(x,y_n) = 0.$$

But U being relatively compact in U, ∂U is compact. The continuous function $y \mapsto k_D(x,y)$ with positive values on ∂U has a positive minimum on U. This contradicts (IV.2.1) and clearly proves our assertion.

Q.E.D.

The following example will show that a similar result does not hold in the infinite dimensional case.

<u>Example</u>. Let $E = \ell_2(\mathbb{N}^\star)$ be the Hilbert space of (unilateral) sequences $x = (x_1, x_2, \dots)$ with norm $\|x\| = (\sum_1^{+\infty} |x_\nu|^2)^{1/2} < \infty$. Setting, for $x \in \ell_2(\mathbb{N}^\star)$, $p(x) = \sup\{\frac{|x_\nu|}{\nu} : \nu=1,2,\dots\}$, we define a norm p on $\ell_2(\mathbb{N}^\star)$ which is continuous, but is not equivalent to the norm $\| \ \|$. Accordingly, the unit ball for p,

$$D = \{x \in \ell_2(\mathbb{N}^\star) : p(x) < 1\}$$

is an unbounded domain in $\ell_2(\mathbb{N}^\star)$.

By Theorem IV.1.13, k_D is a distance and defines an equivalent topology to the topology defined by ρ. However, the latter is not equivalent to the topology defined by $\| \ \|$.

In domains in which either the Carathéodory or the Kobayashi distance defines an equivalent topology to the relative topology, Cartan's Proposition III.2.1 can be considerably improved.

Theorem IV.2.4. *Let* D *be a domain in* E *in which* c_D *(or* k_D*) defines an equivalent topology to the relative topology, and let* $f \in \mathrm{Hol}(D,E)$, *with* $f(D) \subset D$. *If, for some* $x_0 \in D$, $f(x_0) = x_0$ *and* $df(x_0) = \mathrm{Id}$, *then* f *is the identity map.*

Proof. Let $R > 0$ be such that $B(x_0,R) = \{x \in E: \|x-x_0\| < R\} \subset D$. Let S be an open ball with center x_0 and radius $r > 0$ for c_D (or k_D) such that $S \subset B(x_0,R)$. Then S is a bounded domain, and $f_{|S} \in \mathrm{Hol}(S,S)$. Proposition III.2.1 yields the conclusion.

Q.E.D.

Lemma IV.2.5. *Let* D *be as in Theorem* IV.2.4 *and let* $f \in \mathrm{Hol}(D,E)$ *be such that* $f(D) \subset D$ *and* $f(x_0) = x_0$ *at some* $x_0 \in D$. *The spectrum* $\mathrm{Sp}\, df(x_0)$ *of the differential,* $df(x_0)$, *of* f *at* x_0 *is contained in* $\overline{\Delta}$.

Proof. Let S be as before an open ball with center x_0 and radius r for c_D (or k_D). By our hypotheses there exist $R_2 > R_1 > 0$ such that

$$B(x_0,R_1) \subset S \subset B(x_0,R_2).$$

The differential $df^n(x_0)$ of $f^n = f \circ \dots \circ f$ is expressed by

$$df^n(x_0) = df(x_0) \circ \dots \circ df(x_0) = df(x_0)^n.$$

Since $f(S) \subset S$, by Cauchy's inequalities (Proposition II.3.6)

$$\|df(x_0)^n\| \leqslant \frac{1}{\operatorname{dist}(x_0, E\setminus S)} \sup\{\|f(x)\| : x \in S\} \leqslant \frac{R_2}{R_1} .$$

Hence the spectral radius $\rho(df(x_0)) = \lim_{n\to\infty} \|df(x_0)^n\|^{1/n}$ satisfies the inequality

$$\rho(df(x_0)) \leqslant 1.$$

Q.E.D.

Remark. Theorem IV.2.4 and Lemma IV.2.5 are due to H. Cartan and to C. Carathéodory for bounded domains in $\mathbb{C}^n$ and later generalized by W. Kaup and H.H. Wu to hyperbolic manifolds; cf. [Kobayashi, 3, p.75] and references therein. It was also proved by H. Cartan and C. Carathéodory that if D is a bounded domain in $\mathbb{C}^n$ and if all the eigenvalues of $df(x_0)$ have modulus one, then $f \in \operatorname{Aut}(D)$.

The following example shows that a similar result does not hold in the infinite dimensional case.

Let $E = \ell_p(\mathbb{Z})$ $(1 \leqslant p < \infty)$ be the complex Banach space of all bilateral sequences $x = (x_\nu)_{\nu\in\mathbb{Z}}$ with norm

$$\|x\| = \left(\sum_{\nu=-\infty}^{+\infty} |x_\nu|^p\right)^{1/p}.$$

Let (e_ν) be the canonical basis of $\ell_p(\mathbb{Z})$. Choose any $t > 0$, and let T be the bounded linear operator defined on E by

$$Te_\nu = e_{\nu-1} \text{ if } \nu \neq 0 , \quad Te_0 = te_{-1}.$$

It is easily checked that $\|T\| = \max(t,1)$, and that $\operatorname{Sp} T$ is the unit circle (cf. e.g. [Vesentini, 1]). The inverse T^{-1} is defined by

$$T^{-1}e_\nu = e_{\nu+1} \text{ if } \nu \neq -1, \quad T^{-1}\cdot e_{-1} = \frac{1}{t} e_0 ,$$

and moreover $\|T^{-1}\| = \max(\frac{1}{t}, 1)$. Thus, choosing $0 < t < 1$, the corresponding T is a linear (hence holomorphic) map of the unit ball $B \subset E$ into itself such that $\mathrm{Sp}\, T = \{\zeta \in \mathbb{C}: |\zeta| = 1\}$, but which is not an automorphism of B.

We will now discuss briefly completeness for the Kobayashi distance.

<u>Proposition IV.2.5</u>. *Let* D *be a bounded domain in a complex Banach space* E. *If* D *is complete for the Carathéodory distance, then* D *is complete for the Kobayashi distance.*

<u>Proof</u>. Let (x_ν) be a Cauchy sequence for k_D. Then, by (IV.1.4), (x_ν) is a Cauchy sequence for c_D, and hence it converges to some point $x \in D$, for the c_D-topology and therefore also (Theorem IV.2.2) for the relative topology. Since k_D is continuous, then $\lim\limits_{\nu\to\infty} k_D(x_\nu, x) = 0$.

Q.E.D.

<u>Remark</u>. The converse statement does not hold in general. For example, let $\Delta^\star = \{\zeta \in \mathbb{C}: 0 < |\zeta| < 1\}$ be the punctured disc. As Kobayashi has pointed out, $k_{\Delta^\star}$ is a complete distance, whereas c_D is not [Kobayashi, 3, p.60].

The following theorem has been proved by J.-P. Vigué [1, p.279] for the Carathéodory distance.

<u>Theorem IV.2.6</u>. *Let* D *be a bounded domain in a complex Banach space* E. *If* D *is homogeneous (i.e., if the group* $\mathrm{Aut}\, D$ *acts transitively on* D*), then both the Carathéodory and Kobayashi distances are complete.*

<u>Proof</u>. For $x_0 \in D$ and $r > 0$, $B_C(x_0, r)$ and $B(x_0, r)$ will denote as before the open balls with center x_0 and radius r

for c_D and $\| \ \|$ respectively. Let $r>0$ be such that $\overline{B(x_0,r)} \subset\subset D$. By Lemma IV.2.1 there is some $s>0$ for which $B_c(x_0,s) \subset B(x_0,r)$.

Let (x_ν) be a Cauchy sequence for c_D. Let ν_0 be such that $x_\nu \in B_c(x_{\nu_0},s)$ for all $\nu \geqslant \nu_0$. Since D is homogeneous, there is $g \in \mathrm{Aut}(D)$ such that $g(x_{\nu_0}) = x_0$. The sequence $(g(x_\nu))$ is a Cauchy sequence for c_D, and

$$g(x_\nu) \in g(B_c(x_{\nu_0},s)) = B_c(g(x_{\nu_0}),s) = B_c(x_0,s) \subset B(x_0,r)$$

for all $\nu \geqslant \nu_0$. By Theorem IV.2.2, $(g(x_\nu))$ converges to some $x_\infty \in \overline{B(x_0,r)}$ for the norm-topology and therefore for the c_D-topology. Hence (x_ν) converges to $g^{-1}(x_\infty)$ for the c_D-topology.

Proposition IV.2.5 yields completeness for k_D.

Q.E.D.

Proposition IV.2.7. *Let* B *be the open unit ball in a complex Banach space* E, *and suppose that there is a closed ball* K *for* k_B *such that, for every* $x \in B$, $g(x) \in K$ *for some* $g \in \mathrm{Aut}(B)$. *Then* B *is complete for the Kobayashi distance.*

Proof. Let $B_k(x,r) = \{y \in D: k_D(x,y) < r\}$. For any Cauchy sequence (x_ν) for k_B, and for any $r>0$ there is an index ν_0 such that

$$x_\nu \in B_k(x_{\nu_0},r) \qquad \text{for all } \nu \geqslant \nu_0. \tag{IV.2.2}$$

Let $g \in \mathrm{Aut}(B)$ be such that $g(x_{\nu_0}) \in K$. Then

$$g(B_k(x_{\nu_0},r)) = B_k(g(x_{\nu_0}),r) \subset K_r, \tag{IV.2.3}$$

where K_r is the closed set

$$K_r = \{y \in B: k_B(x,y) \leqslant r \quad \text{for some} \quad x \in K\}.$$

Let z_0 and $s \geqslant 0$ be the center and radius of K, and let

$$k = k_B(0,z_0) + s = \omega(0,\|z_0\|) + s.$$

By the triangle inequality

$$k_B(0,x) \leqslant k \qquad \text{for all} \quad x \in K.$$

Let $y \in K_r$, and let $x \in K$ be such that $k_B(x,y) \leqslant r$. Then

$$\omega(0,\|y\|) = k_B(0,y) \leqslant k_B(0,x) + k_B(x,y) \leqslant k + r,$$

i.e.

$$\|y\| \leqslant \frac{e^{2(k+r)}-1}{e^{2(k+r)}+1} .$$

Hence there is an R, with $0 < R < 1$ such that

$$K_r \subset \overline{B(0,R)}.$$

Thus, by (IV.2.2), and (IV.2.3), $g(x_\nu) \subset B(0,R)$ for all $\nu \geqslant \nu_0$.

By Theorem IV.2.2, the sequence $(g(x_\nu))$ converges to an element $x_\infty \in \overline{B(0,R)}$ for the norm-topology and for the k_B-topology. Thus the sequence (x_ν) converges to $g^{-1}(x_\infty)$ for the k_B-topology.

Q.E.D.

Remark. The same kind of argument yields a similar conclusion for the Carathéodory distance. By Proposition IV.2.5, completeness for the Carathéodory distance implies completeness for the Kobayashi distance.

§ 3. Local uniform convergence.

Let U be an open set in a complex normed space E, and $C(U,F)$ be the set of all continuous maps of U into a complex normed space F. Pointwise vector operations define in $C(U,F)$

the structure of a complex vector space, and $Hol(U,F)$ is a subspace of $C(U,F)$.

For any subset $K \subset E$ and any function $f: K \to F$, we set

$$\|f\|_K = \sup\{\|f(x)\|: x \in K\}.$$

For $K \subset U$, the map $f \mapsto \|f\|_K$ is a semi-norm on the subspace of $C(U,F)$ on which $\| \|_K$ takes finite values.

Let $K = \{K_i: i \in I\}$ be a family of subsets $K_i \subset U$, which is closed with respect to finite unions, and, for $f_0 \in$ $\in C(U,F)$, $K \in K$, $\varepsilon > 0$, let

$$U(f_0,K,\varepsilon) = \{f \in C(U,F): \|f-f_0\|_K < \varepsilon\}.$$

When $K \in K$ and $\varepsilon > 0$ vary, the family $\{U(f_0,K,\varepsilon)\}$ defines a fundamental system of neighborhoods in a locally convex topology of $C(U,F)$. This topology is not necessarily Hausdorff.

Taking as K the family of all compact subsets of U, we define in $C(U,F)$ the topology of uniform convergence on compact subsets (compact open topology). Taking as K the family of all (sets consisting of single) points of U we define on $C(U,F)$ the topology of pointwise convergence. The latter is coarser than the compact open topology.

<u>Proposition IV.3.1</u>. *The space* $Hol(U,F)$ *is a closed subspace of* $C(U,F)$ *for the topology of uniform convergence on compact sets.*

<u>Proof</u>. Let f be a point in the closure $\overline{Hol(U,F)}$ of $Hol(U,F)$ in $C(U,F)$. By Theorem II.3.10 it suffices to show that for any choice of $u \in U$, $y \in E$ and any continuous linear form λ on F, the function $\zeta \mapsto \lambda(f(u+\zeta y))$ is holomorphic on the open set $U_{u,y} = \{\zeta \in \mathbb{C}: u+\zeta y \in U\}$ of $\mathbb{C}$, i.e., by Morera's theorem,that, for every simple closed rectifiable curve

Γ, bounding a domain relatively compact in $U_{u,y}$,

$$\text{(IV.3.1)} \qquad \int_\Gamma \lambda(f(u+\zeta y))d\zeta = 0.$$

The image of Γ by the continuous map $\zeta \mapsto u+\zeta y$ is a compact subset $K \subset U$. Let $\varepsilon > 0$, and let $g_\varepsilon \in \mathrm{Hol}(U,F)$ be such that

$$\|f - g_\varepsilon\|_K < \varepsilon .$$

Then for every $\zeta \in \Gamma$

$$|\lambda(f(u+\zeta y)-g_\varepsilon(u+\zeta y))| \leqslant \|\lambda\| \|f(u+\zeta y)-g_\varepsilon(u+\zeta y)\| \leqslant \varepsilon\|\lambda\| .$$

Thus, by Cauchy's integral theorem

$$\begin{aligned} |\int_\Gamma \lambda(f(u+\zeta y))d\zeta| &\leqslant |\int_\Gamma \lambda(f(u+\zeta y)-g_\varepsilon(u+\zeta y))d\zeta| + |\int_\Gamma \lambda(g_\varepsilon(u+\zeta y))d\zeta| = \\ &= |\int_\Gamma \lambda(f(u+\zeta y)-g_\varepsilon(u+\zeta y))d\zeta| \leqslant \\ &\leqslant \int_\Gamma |\lambda(f(u+\zeta y)-g_\varepsilon(u+\zeta y))||d\zeta| \leqslant \varepsilon\|\lambda\| \int_\Gamma |d\zeta| . \end{aligned}$$

Letting $\varepsilon \to 0$, we end up with (IV.3.1).

Q.E.D.

We will now describe another topology on $C(U,F)$.

Let $\mathcal{K}$ be the family of all finite unions of closed balls, for the norm, completely interior to U. The corresponding topology in $C(U,F)$ is called the topology of *local uniform convergence*.

<u>Lemma IV.3.2</u>. *The topology of local uniform convergence is finer than the compact open topology. The two topologies are equivalent if, and only if,* $\dim_{\mathbb{C}} E < \infty$.

<u>Proof</u>. Exercise.

Hence from Proposition IV.3.1 we get

<u>Lemma IV.3.3</u>. $\mathrm{Hol}(U,F)$ *is a closed subspace of* $C(U,F)$ *for*

the topology of local uniform convergence.

In order to investigate the relative topology defined by the local uniform topology in $Hol(U,F)$ we prove first an analogue of Hadamard's three circles theorem.

Theorem IV.3.4. *Let* U *be an open set in the complex Banach space* E *and let* $0<r_1<r_2$. *If* U *contains the annulus* $\{x \in E: r_1 \leq \|x\| \leq r_2\}$, *for any* $f \in Hol(U,F)$ *(*F *complex normed space) the function* $r \mapsto \log \sup\{\|f(x)\|: x \in E, \|x\| = r\}$ *is a convex function of* $\log r$ *for* $r_1 \leq r \leq r_2$.

Proof. The proof is, *mutatis mutandis*, the same as that of the classical Hadamard three circles theorem. Letting

$$M(r) = \sup\{\|f(x)\|: x \in E, \|x\| = r\}$$

we show first that for any $\alpha \in \mathbb{R}$

$$r^{\alpha}M(r) \leq \max(r_1^{\alpha}M(r_1), r_2^{\alpha}M(r_2)) \quad (r_1 \leq r \leq r_2). \tag{IV.3.2}$$

Let us choose any $x \in E$, with $r_1 \leq \|x\| \leq r_2$, and choosing a branch of ζ^{α}, consider the function

$$g_x: \zeta \mapsto \zeta^{\alpha} f(\zeta x)$$

on an open neighborhood of the annulus $A_{r_1 r_2} =$ $= \{\zeta \in \mathbb{C}: \frac{r_1}{\|x\|} \leq |\zeta| \leq \frac{r_2}{\|x\|}\}$, whose interior contains 1.

Since α is not necessarily an integer, g_x is not, in general, holomorphic on a neighborhood of $A_{r_1 r_2}$.

However, for any given point of $A_{r_1 r_2}$, and for any choice of a branch of ζ^{α}, g_x is holomorphic on a neighborhood of that point. Thus the local maximum principle holds, and therefore

$$\|g_x(\zeta)\| \leq \max(\sup\{\|g_x(\zeta)\|: |\zeta| = \frac{r_1}{\|x\|}\}, \quad \sup\{\|g_x(\zeta)\|: |\zeta| = \frac{r_2}{\|x\|}\})$$

for any $\zeta \in \mathbb{C}$ such that $\frac{r_1}{\|x\|} \leqslant |\zeta| \leqslant \frac{r_2}{\|x\|}$. In other words

$$|\zeta|^\alpha \|f(\zeta x)\| \leqslant \max\left(\left(\frac{r_1}{\|x\|}\right)^\alpha \sup\{\|f(\zeta x)\| : |\zeta| = \frac{r_1}{\|x\|}\}\right.,$$

$$\left.\left(\frac{r_2}{\|x\|}\right)^\alpha \sup\{\|f(\zeta x)\| : |\zeta| = \frac{r_2}{\|x\|}\}\right)$$

for $\frac{r_1}{\|x\|} \leqslant |\zeta| \leqslant \frac{r_2}{\|x\|}$, and therefore

$$\|\zeta x\|^\alpha \|f(\zeta x)\| \leqslant \max(r_1^\alpha M(r_1), r_2^\alpha M(r_2)) \qquad \left(\frac{r_1}{\|x\|} \leqslant |\zeta| \leqslant \frac{r_2}{\|x\|}\right).$$

In particular, for $\zeta=1$,

$$\|x\|^\alpha \|f(x)\| \leqslant \max(r_1^\alpha M(r_1), r_2^\alpha M(r_2))$$

for all $x \in E$, with $r_1 \leqslant \|x\| \leqslant r_2$, and in conclusion we have established (IV.3.2).

Let us choose now α in such a way that

$$r_1^\alpha M(r_1) = r_2^\alpha M(r_2)$$

i.e.

$$\alpha \log \frac{r_1}{r_2} = \log \frac{M(r_2)}{M(r_1)} .$$

Then

$$r^\alpha M(r) \leqslant r_1^\alpha M(r_1)$$

and therefore

$$\log M(r) \leqslant \log M(r_1) + \alpha \log \frac{r_1}{r} . \tag{IV.3.3}$$

Writing

$$\log r = t \log r_1 + (1-t) \log r_2 \qquad (0 \leqslant t \leqslant 1) \tag{IV.3.4}$$

we have

$$\log r_1 - \log r = (1-t) \log \frac{t_2}{r_2} = \frac{1-t}{\alpha} \log \frac{M(r_2)}{M(r_1)}$$

i.e., by (IV.3.4)

$$\log M(r) \leqslant \log M(r_1) + (1-t) \log \frac{M(r_2)}{M(r_1)} = t \log M(r_1) + (1-t) \log M(r_2).$$

Q.E.D.

The last inequality can be written also as

(IV.3.5) $$M(r) \leqslant M(r_1)^t M(r_2)^{1-t} \qquad (0 \leqslant t \leqslant 1).$$

Since, by (IV.3.4)

$$t = \frac{\log r_2 - \log r}{\log r_2 - \log r_1} = \log \frac{r_2}{r} \Big/ \log \frac{r_2}{r_1},$$

$$1-t = \log \frac{r}{r_1} \Big/ \log \frac{r_2}{r_1}$$

then (IV.3.5) takes the classical aspect of Hadamard's three circles theorem

$$M(r)^{\log\frac{r_2}{r_1}} \leqslant M(r_1)^{\log\frac{r_2}{r}} M(r_2)^{\log\frac{r}{r_1}} \qquad (r_1 \leqslant r \leqslant r_2).$$

The following lemma is a trivial consequence of Theorem IV.3.4.

<u>Lemma IV.3.5</u>. *Let* B^1, B^2, B^3 *be three concentric closed balls contained in the open set* U, *with radii* r_1, r_2, r_3, $0 < r_1 \leqslant \leqslant r_2 < r_3$. *Given* $\varepsilon_2 > 0$, $\varepsilon_3 > 0$ *there exists* $\varepsilon_1 > 0$ *such that, whenever* $f \in \mathrm{Hol}(U,F)$ *satisfies the conditions*

$\|f(x)\| \leqslant \varepsilon_3$ *for all* $x \in B^3$, $\|f(x)\| \leqslant \varepsilon_1$ *for all* $x \in B^1$,

then

$\|f(x)\| \leqslant \varepsilon_2$ *for all* $x \in B^2$.

<u>Lemma IV.3.6</u>. *Let* D *be a domain in* E. *For any choice of* x,y *in* D *there exists a finite sequence* $(x_0, x_1, \dots, x_n)$ *of points and a finite sequence* $(r_0, \dots, r_{n-1})$ *of positive numbers such that:* $x_0 = x$, $x_n = y$, $x_i \in B(x_{i-1}, r_{i-1})$ $(i=1,\dots,n)$, $\overline{B(x_i, r_i)} \subset\subset D$ $(i=0,1,\dots,n-1)$.

<u>Proof</u>. Exercise (Hint: D, being connected and locally arcwise connected, is arcwise connected).

For $M > 0$, let $\mathrm{Hol}(D,F)_M = \{f \in \mathrm{Hol}(D,F) : \|f(x)\| \leqslant M$ for all $x \in D\}$.

Proposition IV.3.7. *Let* D *be a (proper) domain in* E. *Let* B^0, B^1 *be two closed balls completely interior to* D. *For any* $\varepsilon > 0$ *there exists* $\eta > 0$ *such that whenever* $f \in Hol(D,F)_M$ *satisfies the condition* $\|f(x)\| \leqslant \eta$ *for all* $x \in B^0$, *then* $\|f(x)\| \leqslant \varepsilon$ *for all* $x \in B^1$.

Proof. Let B be a closed ball, $B \subset C$. Let $x \in \mathring{B}$, and let B', B'' be two closed balls with center x, such that $B' \subset B$, $B' \subset B'' \subset\subset D$. By Lemma IV.3.5, given $\delta > 0$, there exists $\eta > 0$ such that whenever $f \in Hol(D,F)_M$ satisfies the condition $\|f(x)\| \leqslant \eta$ for all $x \in B'$, then $\|f(x)\| \leqslant \delta$ for all $x \in B''$. To prove the proposition we need only apply Lemma IV.3.6, and then the above argument a finite number of times.

Q.E.D.

Remark. In view of the above proposition, the topology induced on $Hol(D,F)_M$ by the topology of local uniform convergence can be described as follows. Choose a closed ball $B \subset\subset D$. For any $f_0 \in Hol(D,F)_M$ and any $\varepsilon > 0$, let

$$W(f_0,B,\varepsilon) = \{f \in Hol(D,F)_M : \|f-f_0\|_B < \varepsilon\}.$$

Then $\{W(f_0,B,\varepsilon) : \varepsilon > 0\}$ is a fundamental system of neighborhoods of f_0. Note that B may be kept fixed and that f_0 has a countable fundamental system of neighborhoods.

We will now consider the group $Aut(D)$ of all holomorphic automorphisms of D, endowed with the relative topology defined by $Hol(D,E)$. We shall show that, if D is bounded, then $Aut(D)$ is a topological group acting continuously on D.

Let D be a metric space with a distance function d, and let G be the set of all isometries of D into D.

We consider on G the topology of uniform convergence on finite unions of balls. For $g_0 \in G$, a fundamental system of neighborhoods of g_0 is the following. Let $\varepsilon > 0$, let $B_1, \dots B_n$ be closed balls in D, and let

(IV.3.6) $$V = \bar{B}_1 \cup \dots \cup \bar{B}_n.$$

The set

$$U(g_0, V, \varepsilon) = \{g \in G: d(g_0(x), g(x)) < \varepsilon \quad \text{for all} \quad x \in V\}$$

describes, when ε and V vary, a fundamental system of neighborhoods of g_0.

Clearly this topology is Hausdorff.

Theorem IV.3.8. *The group* G *is a topological group and the natural mapping* $G \times D \to D$ *is continuous.*

Proof. 1) We prove that the product $G \times G \to G$ is continuous.

For $g_1^0, g_2^0 \in G$, any neighborhood of $g_1^0 \circ g_2^0$ contains a neighborhood of the form

$$U(g_1^0 \circ g_2^0, V, \varepsilon) = \{g \in G: d(g_1^0 \circ g_2^0(x), g(x)) < \varepsilon, \quad x \in V\},$$

where $\varepsilon > 0$ and V is a finite union of closed balls, given by (IV.3.6). Since g_2^0 is an isometry, $g_2^0(V)$ is the union of the closed balls $g_2^0(B_j) \subset D$ $(j=1, \dots, n)$. For all $g_1 \in U(g_1^0, g_2^0(V), \frac{\varepsilon}{2})$, $g_2 \in U(g_2^0, V, \frac{\varepsilon}{2})$, and any $x \in V$

$$d(g_1 \circ g_2(x), g_1^0 \circ g_2^0(x)) \leq d(g_1 \circ g_2(x), g_1 \circ g_2^0(x)) + d(g_1 \circ g_2^0(x), g_1^0 \circ g_2^0(x))$$
$$= d(g_2(x), g_2^0(x)) + d(g_1 \circ g_2^0(x), g_1^0 \circ g_2^0(x)) < \frac{\varepsilon}{2} + \frac{\varepsilon}{2} = \varepsilon.$$

2) For any $g_0 \in G$, and any neighborhood

$$U(g_0^{-1}, V, \varepsilon),$$

for all $g \in U(g_0, g_0^{-1}(V), \varepsilon)$ and any $x \in V$, which we can express as $x = g_0(y)$ with $y = g_0^{-1}(x) \in g_0^{-1}(V)$, we have

$$d(g^{-1}(x), g_0^{-1}(x)) = d(g^{-1}(g_0(y)), y) = d(g_0(y), g(y)) < \varepsilon.$$

3) We are left to prove that the map $G \times D \to D$ is continuous. Let $g_0 \in G$, $x_0 \in D$; choose any $\varepsilon > 0$ and let $B(x_0, \varepsilon) = \{x \in D: d(x_0, x) < \varepsilon\}$ and $U = U(g_0, \{x_0\}, \frac{\varepsilon}{2}) = \{g \in G: d(g(x_0), g_0(x_0)) < \frac{\varepsilon}{2}\}$. Then for all $x \in B(x_0, \frac{\varepsilon}{2})$ and all $g \in U$,

$$d(g(x), g_0(x_0)) \leqslant d(g(x), g(x_0)) + d(g(x_0), g_0(x_0)) =$$
$$= d(x, x_0) + d(g(x_0), g_0(x_0)) < \frac{\varepsilon}{2} + \frac{\varepsilon}{2} = \varepsilon.$$

Q.E.D.

Now, let D be a bounded domain in a complex Banach space E, and let d be the Kobayashi distance on D (the Carathéodory distance would also do).

The group $\mathrm{Aut}(D)$ of all holomorphic automorphisms of D is a subgroup of G; hence it is a topological group for the relative topology in G.

In view of the above Remark and of Lemmas IV.1.10 and IV.2.1 we have

Lemma IV.3.10. *The relative topology in* $\mathrm{Aut}(D)$ *is equivalent to the topology of local uniform convergence.*

Hence we have

Corollary IV.3.11. *If* D *is a bounded domain in* E *the group* $\mathrm{Aut}(D)$ *is a topological group acting continuously on* D.

Let $\mathrm{Iso}_k(D)$ (or $\mathrm{Iso}_c(D)$) be the semigroup of all holomorphic maps of D into D, which are isometric for the Kobayashi distance k_D (or, respectively, the Carathéodory distance c_D). Proposition IV.3.9 and the previous Remark yield

Proposition IV.3.12. *Both* $\mathrm{Iso}_k(D)$ *and* $\mathrm{Iso}_c(D)$ *with the re-*

lative topology in Hol(D,D) *are topological semi-groups acting continuously on* D.

In chapter VI we shall describe in detail the unit ball of a complex Hilbert space, its Carathéodory and Kobayashi distances and its automorphisms. Now we will consider briefly another relevant example.

Let X be a compact Hausdorff space and let $C(X)$ be the algebra of all complex valued continuous functions on X; pointwise composition and uniform norm $\| \|_\infty$ make $C(X)$ a Banach algebra with identity.

Let B be the open unit ball of $C(X)$ and let $f_0 \in B$. For any $f \in B$ we denote by $F_{f_0}(f)$ the continuous function defined on X by

$$(F_{f_0}(f))(x) = \frac{f(x) + f_0(x)}{1 + \overline{f_0(x)} f(x)}$$

For any fixed $x \in X$, the map $f(x) \mapsto (F_{f_0}(f))_x$ is a Moebius transformation of the unit disc Δ. Thus F_{f_0} defines a map of B into itself. A direct computation shows that F_f is bijective, being $F_{f_0}^{-1} = F_{-f_0}$. We prove now that

$$F_{f_0} \in \mathrm{Hol}(B,B) \tag{IV.3.7}$$

so that in conclusion $F_{f_0} \in \mathrm{Aut}(B)$.

First of all, for any $x \in X$, let $\delta_x : f \mapsto f(x)$ be the evaluation map at x. The linear form δ_x is continuous; the set $\{\lambda\delta_x : x \in X, \lambda \in \mathbb{C}, |\lambda| = 1\}$ is the set of all extreme points of the closure of the unit ball of the dual space $C(X)'$ of $C(X)$ (cf. e.g. [Dunford-Schwartz, 1, pp. 441-442]). Hence the linear forms δ_x $(x \in X)$ span the entire space $C(X)'$. Let $f_1, f_2 \in B$ and let U_{f_1,f_2} be the open set in $\mathbb{C}$

$$U_{f_1,f_2} = \{\zeta \in \mathbb{C} : f_1 + \zeta f_2 \in \mathbb{C}\}.$$

For any fixed $x \in X$ the function $\zeta \mapsto \delta_x(F_{f_0}(f_1+\zeta f_2))$ is expressed on U_{f_1,f_2} by

$$\delta_x(F_{f_0}(f_1+\zeta f_2)) = \frac{f_1(x)+\zeta f_2(x)+f_0(x)}{1-\overline{f_0(x)}(f_1(x)+\zeta f_2(x))},$$

and is therefore holomorphic. Theorem II.3.10 proves then (IV.3.7).

Since $F_{f_0}(f_0) = 0$, the group $\mathrm{Aut}(B)$ acts transitively on B, i.e. B *is homogeneous.*

We will now determine the entire group $\mathrm{Aut}(B)$. Let $F \in \mathrm{Aut}(B)$ and let $f_0 \in F^{-1}(0)$. The holomorphic automorphism of B $T = F \circ F_f^{-1}$ leaves the origin fixed. By Proposition III.2.2 T is (the restriction to B) of a linear isomorphism of $C(X)$, i.e. T is a linear isometry of $C(X)$ onto $C(X)$.

According to the Banach-Stone theorem (cf. e.g. [Dunford-Schwartz, 1, pp. 442-443]) there exist a unique homeomorphism τ of X onto itself and a unique continuous function α on X such that $|\alpha(x)| = 1$ and

$$(Tf)(x) = \alpha(x) f(\tau(x))$$

for all $x \in X$ and all $f \in C(X)$. In conclusion we have proved

Theorem IV.3.13. *For every* $F \in \mathrm{Aut}(B)$ *there exist a unique function* $\alpha \in C(X)$, *and a unique homeomorphism* τ *of* X *onto itself such that* $|\alpha(x)| = 1$ *for all* $x \in X$ *and, setting* $f_0 = F^{-1}(0)$,

$$F(f)(x) = \alpha(x) \frac{f(\tau(x)) - f_0(\tau(x))}{1-\overline{f_0(\tau(x))} f(\tau(x))}.$$

Note that F is continuous on $\overline{B}$.

We compute now invariant distances on B. Since B is homogeneous, by Theorem IV.1.8, the Carathéodory and Kobayashi distances on B coincide. For $f_1, f_2 \in B$, then

$$k_B(f_1,f_2) = c_B(f_1,f_2) = c_B(F_{-f_1}(f_1),F_{-f_1}(f_2)) = c_B(0,F_{-f_1}(f_2)) = \\ = \omega(0,\|F_{-f_1}(f_2)\|)$$

i.e.

$$k_B(f_1,f_2) = c_B(f_1,f_2) = \omega\left(0, \left\|\frac{f_1-f_2}{1-\overline{f_1}f_2}\right\|_\infty\right).$$

Example. The following example, due to L. Stachò, shows that there are automorphisms of B with no fixed point in $\overline{B}$. Let $X = \overline{\Delta}$, and let $f_0 : \zeta \mapsto \frac{1}{2}\zeta$. The map $F \in \mathrm{Aut}(B)$ defined by

$$F(f)(\zeta) = \frac{f(\zeta)-f_0(\zeta)}{1-\overline{f_0(\zeta)}f(\zeta)} = \frac{f(\zeta) - \frac{1}{2}\zeta}{1 - \frac{1}{2}\overline{\zeta}\,f(\zeta)}$$

is the restriction to B of a holomorphic map of the ball $B_2 = \{f \in C(\overline{\Delta}) : \|f\|_\infty < 2\}$ into $C(\overline{\Delta})$. Let $u \in B_2$ be a fixed point of this map. Then

$$u(\zeta) - \frac{1}{2}\zeta = u(\zeta) - \frac{1}{2}\overline{\zeta}\,\overline{u(\zeta)} \qquad \text{for all} \quad \zeta \in \overline{\Delta}.$$

i.e.

$$u(\zeta) = \frac{\zeta}{\overline{\zeta}} \qquad \text{for all} \quad \zeta \in \overline{\Delta}.$$

But this is absurd, since the function $\zeta \mapsto \frac{\zeta}{\overline{\zeta}}$ is not continuous at the point 0.

Notes.

The Kobayashi distance was introduced by S. Kobayashi in [1,2,3]. Our presentation follows closely Kobayashi's description. Many of the proofs in § 1 are adapted, *mutatis mutandis*, from [3] where only the finite dimensional case is considered.

The Carathéodory distance was introduced by C. Carathéodory in [1,2,3] for a domain D in $\mathbb{C}^2$. The fact that $c_D(x,y) < \infty$

for all $x,y \in D$, was established by Carathéodory by an argument using the theory of normal families (which actually shows that the supremum appearing in (IV.1.3) is a maximum). Inequality (IV.1.4), due to Kobayashi, avoids the use of normal families.

The Carathéodory distance for domains in Banach spaces was investigated by J.-P. Vigué in [1, 278-280], with applications to domains of holomorphy. Theorems IV.2.2, IV.3.5, and IV.2.6 are due to J.-P. Vigué [1], while Lemma IV.2.1 and Theorem IV.2.4 appear in [Earle-Hamilton, 1]. Proposition IV.2.3 is due to T.J. Barth [1].

For an extension of some of these results to domains in locally convex spaces cf. [Vesentini, 4].

As for the explicit construction of Aut D for specific examples it is worth pointing out that the structure of Aut(D) is still largely unknown for bounded finitely dimensional homogeneous domains. In fact, if D is a bounded *symmetric* domain in $\mathbb{C}^n$, the construction of Aut D was carried out by C.L.Siegel [1], H. Klingen [1,2], U. Hirzebruch [1]; cf. [C.L. Siegel, 3] for relevant bibliographical references. Some of these results have been extended to bounded symmetric domains in (infinite dimensional) complex Hilbert spaces; cf. [R.S. Phillips, 1], [Greenfield and Wallach, 1],[L.A. Harris, 2],[J.P.Vigué,1 and 2].

In [Vesentini, 3] it was shown that, given any measure space (M,Ξ,μ) on a set M, the open unit ball of $L^1(M,\Xi,\mu)$ is highly non-homogeneous whenever $\dim_{\mathbb{C}} L^1(M,\Xi,\mu) > 1$, in the sense that every holomorphic automorphism of B is (the restriction to B of) a linear isometry of $L^1(M,\Xi,\mu)$. For an extension of this result to $L^p(M,\Xi,\mu)$ $(1 \leq p < \infty,\ p \neq 2)$, cf. [Braun-Kaup-Upmeier, 1] where also the case of the unit balls of Hardy spaces is considered. These results can be viewed also as fixed point theorems. For general results about fixed points for automorphisms of the unit ball in a Banach lattice of type L cf. [Stachò, 1].

CHAPTER V

INVARIANT DIFFERENTIAL METRICS

§ 1. The Kobayashi and Carathéodory differential metrics.

For $\zeta \in \Delta$, $\tau \in \mathbb{C}$ the "length" of the vector τ for the Poincaré metric at ζ is the number $<\tau>_\zeta$ expressed by

$$<\tau>_\zeta = \frac{|\tau|}{1 - |\zeta|^2} .$$

As usual, let D be a domain in a complex normed space E.

Lemma V.1.1. *For all* $v \in E$, $x \in D$, $\zeta_0 \in \Delta$ *there exist* $h \in Hol(\Delta, D)$ *and* $\tau \in \mathbb{C}$ *such that* $h(\zeta_0) = x$, $dh(\zeta_0)\tau = v$.

Proof. Let $r > 0$ be such that $B(x,r) \subset D$. If $v \neq 0$, the holomorphic function $k: \zeta \mapsto x + \frac{r\zeta}{\|v\|} v$ maps Δ into $B(x,r)$. Furthermore $k(0) = x$, $k'(0) = \frac{r}{\|v\|} v$. Thus the function $h \in Hol(\Delta, D)$ defined by $h(\zeta) = k\left(\frac{\zeta - \zeta_0}{1 - \bar{\zeta}_0 \zeta}\right)$ satisfies all the requirements of the lemma, for a suitable value of the complex number τ.

Q.E.D.

The *Kobayashi differential metric* is the function $\kappa_D: D \times E \to \mathbb{R}_+$ defined on (x,v) by

$$\text{(V.1.1)} \qquad \kappa_D(x;v) = \inf <\tau>_\zeta ,$$

where the infimum is taken over all $\tau \in \mathbb{C}$, $\zeta \in \Delta$, $h \in Hol(\Delta, D)$ such that

$$(V.1.2) \qquad h(\zeta) = x, \qquad d h(\zeta)\tau = v.$$

Since the group $\text{Aut}(\Delta)$ is transitive on Δ, and its elements are isometries for the Poincaré metric, then $\zeta \in \Delta$ can be chosen arbitrarily and kept fixed in the above definition. Choosing in particular $\zeta = 0$, (V.1.1) reads now

$$\kappa_D(x;v) = \inf\{|\tau| : \tau \in \mathbb{C},\ h \in \text{Hol}(\Delta,D),\ h(0) = x,\ dh(0)\tau = v\}.$$

Let $h \in \text{Hol}(\Delta,D)$ be as in (V.1.2) and let $g \in \text{Hol}(D,\Delta)$. Then, denoting by $dg(x)$ the differential of g at the point x,

$$dg(x)(v) = dg(h(\zeta))dh(\zeta)(\tau) = d(g\circ h)(\tau).$$

Since $g\circ h \in \text{Hol}(\Delta,\Delta)$, by the Schwarz-Pick lemma,

$$\langle dg(x)(v)\rangle_{g(x)} \leqslant \langle\tau\rangle_\zeta .$$

This inequality holds for any $h \in \text{Hol}(\Delta,D)$ satisfying (V.1.2). Then

$$\langle dg(x)(v)\rangle_{g(x)} \leqslant \kappa_D(x;v).$$

Thus the number $\gamma_D(x;v) \in \mathbb{R}_+$ defined by

$$\gamma_D(x;v) = \sup\{\langle dg(x)v\rangle_{g(x)} : g \in \text{Hol}(D,\Delta)\},$$

satisfies the inequality

$$(V.1.3) \qquad \gamma_D(x;v) \leqslant \kappa_D(x;v) \qquad (x \in D,\quad v \in E),$$

and therefore is finite. The function $\gamma_D: D \times E \to \mathbb{R}_+$ is called the *Carathéodory differential metric* on D.

For any $a \in \mathbb{C}$

$$(V.1.4) \qquad \kappa_D(x;av) = |a|\kappa_D(x;v), \qquad \gamma_D(x;av) = |a|\gamma_D(x;v).$$

Furthermore, for $v_1, v_2 \in E$

$$(V.1.5) \qquad \gamma_D(x;v_1+v_2) \leqslant \gamma_D(x;v_1) + \gamma_D(x;v_2).$$

Let D_1 be a domain in a complex normed space E_1. The following proposition is an obvious consequence of the definitions.

<u>Proposition V.1.2</u>. *For all* $F \in \mathrm{Hol}(D, D_1)$, $x \in D$, $v \in E$,

$$\kappa_{D_1}(F(x);dF(x)v) \leqslant \kappa_D(x;v),\quad \gamma_{D_1}(F(x);dF(x)v) \leqslant \gamma_D(x;v).$$

In particular, for $F \in \mathrm{Aut}(D)$,

$$\kappa_D(F(x);dF(x)v) = \kappa_D(x;v),\quad \gamma_D(F(x);dF(x)v) = \gamma_D(x;v).$$

Furthermore the Kobayashi differential metric is the "largest" differential metric for which every holomorphic map $h\colon \Delta \to D$ is a contraction. More precisely we have, as an immediate consequence of the definition,

<u>Proposition V.1.3</u>. *Let* $\sigma\colon D \times E \to \mathbb{R}$ *be a function such that*

$$\sigma(x;av) = |a|\sigma(x;v) \quad \textit{for all} \quad x \in D,\ a \in \mathbb{C},\ v \in E, \tag{V.1.6}$$

and that

$$\sigma(x;dh(0)1) \leqslant 1$$

for all $h \in \mathrm{Hol}(\Delta,D)$, *with* $h(0) = x$. *Then*

$$\sigma(x;v) \leqslant \kappa_D(x;v) \quad \textit{for all} \quad x \in D,\ v \in E.$$

Similarly we have

<u>Proposition V.1.4</u>. *Let* $\sigma\colon D \times E \to \mathbb{R}$ *be a function satisfying* (V.1.6) *and such that*

$$\sigma(x;v) \geqslant |df(x)v|$$

for all $f \in \mathrm{Hol}(D,\Delta)$, *with* $f(x) = 0$. *Then*

$$\sigma(x;v) \geqslant \gamma_D(x;v) \quad \textit{for all} \quad x \in D,\ v \in E.$$

Let $E = \mathbb{C}$, $D = \Delta$. Taking $g\colon \zeta \mapsto \zeta$ in the definition of γ_Δ, we have

$$\gamma_\Delta(\zeta;\tau) \geqslant \langle\tau\rangle_\zeta .$$

Being $dg(\zeta)\tau = \tau$, we have also

$$\kappa_\Delta(\zeta;\tau) \leqslant \langle\tau\rangle_\zeta .$$

Thus, by (V.1.3),

$$\text{(V.1.7)} \qquad \gamma_\Delta(\zeta;\tau) = \kappa_\Delta(\zeta;\tau) = \langle\tau\rangle_\zeta \qquad (\zeta \in \Delta, \quad \tau \in \mathbb{C}).$$

More in general, for $\Delta_R = \{\zeta \in \mathbb{C}: |\zeta| < R\}$,

$$\gamma_{\Delta_R}(\zeta;\tau) = \kappa_{\Delta_R}(\zeta;\tau) = \left\langle\frac{\tau}{R}\right\rangle_{\frac{\zeta}{R}} = \frac{R|\tau|}{R^2-|\zeta|^2} \qquad (\zeta \in \Delta_R , \quad \tau \in \mathbb{C}).$$

<u>Lemma V.1.5</u>. *Let* p *be a continuous semi-norm on* E, *and let* $B_p = \{x \in E: p(x) < 1\}$. *Then*

$$\gamma_{B_p}(0;v) = \kappa_{B_p}(0;v) = p(v) \qquad (v \in E).$$

<u>Proof</u>. Suppose first that $p(v) > 0$, and let λ be a continuous linear form on E such that $|\lambda(y)| \leqslant p(y)$ for all $y \in E$, and $\lambda(v) = p(v)$. Then $\lambda \in \mathrm{Hol}(B_p ,\Delta)$. If $h \in \mathrm{Hol}(\Delta,B_p)$ is defined by $h(\zeta) = \frac{\zeta}{p(v)}v$, then Lemma V.1.5 follows from the sequence of inequalities

$$p(v) = \gamma_\Delta(0;p(v)) = \gamma_\Delta(\lambda(0);\lambda(v)) = \gamma_\Delta(\lambda(0);d\lambda(0)v) \leqslant \gamma_{B_p}(0;v) \leqslant$$
$$\leqslant \kappa_{B_p}(0;v) = \kappa_{B_p}(h(0);dh(0)p(v)) \leqslant \kappa_\Delta(0;p(v)) = p(v).$$

If $p(v) = 0$, let $t > 1$ and let $h_t \in \mathrm{Hol}(\mathbb{C},B_p)$ be defined by $h_t(\zeta) = t\zeta v$. Then

$$\gamma_{B_p}(0;v) \leqslant \kappa_{B_p}(0;v) = \kappa_{B_p}(h_t(0);\ dh_t(0)\tfrac{1}{t}) \leqslant \kappa_\Delta(0;\tfrac{1}{t}) = \frac{1}{t} .$$

Letting $t \to \infty$, the conclusion follows.

Q.E.D.

More in general, for any $x \in E$, $r > 0$, let $B_p(x,r) =$

$= \{y \in E\colon p(x-y) < r\}$. Then

$$\gamma_{B_p(x,r)}(x;v) = \kappa_{B_p(x,r)}(x;v) = \frac{p(v)}{r}.$$

Choosing $p = 0$, we obtain

$$\gamma_E(x;v) = \kappa_E(x;v) = 0.$$

Examples. 1. Let $E = \mathbb{C}^2$ and $\|(\zeta^1,\zeta^2)\| = \max(|\zeta^1|,|\zeta^2|)$. Then

$$\gamma_{\Delta\times\Delta}((0,0);(\tau^1,\tau^2)) = \kappa_{\Delta\times\Delta}((0,0);(\tau^1,\tau^2)) = \max(|\tau^1|,|\tau^2|).$$

For $(\zeta^1,\zeta^2) \in \Delta\times\Delta$, the automorphism of $\Delta\times\Delta$ $(\eta^1,\eta^2) \mapsto \mapsto \left(\frac{\eta^1-\zeta^1}{1-\overline{\zeta^1}\eta^1}, \frac{\eta^2-\zeta^2}{1-\overline{\zeta^2}\eta^2}\right)$ maps (ζ^1,ζ^2) into $(0,0)$. Thus

$$\gamma_{\Delta\times\Delta}((\zeta^1,\zeta^2);(\tau^1,\tau^2)) = \gamma_{\Delta\times\Delta}\left((0,0);\left(\frac{\tau^1}{1-|\zeta^1|^2}, \frac{\tau^2}{1-|\zeta^2|^2}\right)\right) =$$

$$= \max\left(\frac{|\tau^1|}{1-|\zeta^1|^2}, \frac{|\tau^2|}{1-|\zeta^2|^2}\right) = \max(<\tau^1>_{\zeta^1},<\tau^2>_{\zeta^2})$$

and also

$$\kappa_{\Delta\times\Delta}((\zeta^1,\zeta^2);(\tau^1,\tau^2)) = \max(<\tau^1>_{\zeta^1},<\tau^2>_{\zeta^2}).$$

Similarly, we have

$$\gamma_{\Delta_{R_1}\times\Delta_{R_2}}((\zeta_1,\zeta_2),(\tau_1,\tau_2)) = \kappa_{\Delta_{R_1}\times\Delta_{R_2}}((\zeta_1,\zeta_2),(\tau_1,\tau_2)) =$$

$$= \max\{<\frac{\tau_1}{R_1}>_{\zeta_1/R_1},<\frac{\tau_2}{R_2}>_{\zeta_2/R_2}\}.$$

Proposition V.1.6. *Let* D_1 *and* D_2 *be two domains in two complex normed spaces* E_1 *and* E_2. *Then for* $x_j \in D_j$, $v_j \in E_j$ $(j=1,2)$ *we have*

$$\kappa_{D_1\times D_2}((x_1,x_2);(v_1,v_2)) = \max(\kappa_{D_1}(x_1;v_1),\kappa_{D_2}(x_2;v_2)).$$

Proof. The canonical maps $D_1 \times D_2 \to D_1$, $D_1 \times D_2 \to D_2$ are holomorphic. Thus, by Proposition V.1.2,

$$\kappa_{D_1\times D_2}((x_1,x_2);(v_1,v_2)) \geqslant \kappa_{D_j}(x_j;v_j) \qquad (j=1,2).$$

Suppose that

$$\kappa_{D_1 \times D_2}((x_1,x_2);(v_1,v_2)) > \max(\kappa_{D_1}(x_1;v_1),\kappa_{D_2}(x_2;v_2)),$$

and let $f_j \in \mathrm{Hol}(\Delta,D_j)$, $\tau_j \in \mathbb{C}$ be such that

$$f_j(0) = x_j \;, \quad df_j(0)\tau_j = v_j \;,$$

where $\tau_j \in \mathbb{C}$ is such that

$$\kappa_{D_1 \times D_2}((x_1,x_2);(v_1,v_2)) > |\tau_j| \geqslant \kappa_{D_j}(x_j;v_j) \quad (j=1,2). \tag{V.1.8}$$

The function $f\colon (\zeta_1,\zeta_2) \mapsto (f_1(\zeta_1),f_2(\zeta_2))$ belongs to $\mathrm{Hol}(\Delta\times\Delta, D_1\times D_2)$, and

$$f(0,0) = (x_1,x_2), \quad df(0,0)(\tau_1,\tau_2) = (v_1,v_2).$$

Thus, by Proposition V.1.2 and the previous example, we have

$$\kappa_{D_1 \times D_2}((x_1,x_2);(v_1,v_2)) = \kappa_{D_1 \times D_2}(f(0,0);df(0,0)(\tau_1,\tau_2)) \leqslant$$

$$\leqslant \kappa_{\Delta\times\Delta}((0,0);(\tau_1,\tau_2)) = \max(|\tau_1|,|\tau_2|).$$

But this contradicts (V.1.8).

Q.E.D.

<u>Example 2</u>. *Let* p_1 *and* p_2 *be continuous semi-norms on two complex normed spaces* E_1 *and* E_2. *Then, for* $R_1 > 0$, $R_2 > 0$, $v_1 \in E_1$, $v_2 \in E_2$,

$$\gamma_{B_{p_1}(0,R_1)\times B_{p_2}(0,R_2)}((0,0);(v_1,v_2)) = \max\{\gamma_{B_{p_1}(0,R_1)}(0;v_1),\gamma_{B_{p_2}(0,R_2)}(0;v_2)\},$$

$$\kappa_{B_{p_1}(0,R_1)\times B_{p_2}(0,R_2)}((0,0);(v_1,v_2)) =$$

$$= \max\{\kappa_{B_{p_1}(0,R_1)}(0;v_1),\ \kappa_{B_{p_2}(0,R_2)}(0;v_2)\}.$$

In fact, the function $p = \max\{p_1,p_2\}$ is a continuous semi-norm on $E_1 \times E_2$ for the product topology, and the result follows from Lemma V.1.5.

The following lemma implies the continuity of $\kappa_{B_p(x,R)}$ at $(x,0)$.

Lemma V.1.7. *For any* $\varepsilon > 0$, *there is a neighborhood* U *of* $(x,0)$ *in* $B_p(x,R) \times E$ *such that*

$$\kappa_{B_p(x,R)}(y;v) < \varepsilon$$

for all $(y,v) \in U$.

Proof. Let $U = B_p(x,\frac{R}{2}) \times B_p(0,\frac{R\varepsilon}{4})$. For $y \in B_p(x,\frac{R}{2})$, $v \in B_p(0,\frac{R\varepsilon}{4})$ the function $f \in \mathrm{Hol}(\Delta,E)$ defined by $f(\zeta) = y + \frac{2}{\varepsilon}\zeta v$ maps Δ into $B_p(x,R)$. Since $f(0) = y$, $df(0)\frac{\varepsilon}{2} = v$, by Proposition V.1.2,

$$\kappa_{B_p(x,R)}(y;v) \leqslant \kappa_\Delta(0;\tfrac{\varepsilon}{2}) = \frac{\varepsilon}{2} < \varepsilon .$$

Q.E.D.

For $x \in D$, let $r > 0$ be such that $B(x,r) \subset D$. Then

$$\text{(V.1.9)} \qquad \gamma_D(x;v) \leqslant \kappa_D(x;v) \leqslant \kappa_{B(x,r)}(x;v) = \frac{\|v\|}{r} \quad \text{for all } v \in E.$$

Remark. If there is a continuous semi-norm p on E such that $B_p(x,r) \subset D$ for some $r > 0$, then $\gamma_D(x;v) = \kappa_D(x;v) = 0$ for all $v \in E$ for which $p(v) = 0$.

Inequalities (V.1.9) and a trivial compactness argument yield

Lemma V.1.8. *For every compact subset* $K \subset D$ *there is a positive constant* k *such that*

$$\gamma_D(x,v) \leqslant \kappa_D(x,v) \leqslant k\|v\|$$

for all $x \in K$ *and all* $v \in E$.

Proposition V.1.9. *Let* D *be a hyperbolic domain. Then for all* $x \in D$ *and all* $v \in E\setminus\{0\}$, $\kappa_D(x;v) > 0$.

Proof. Suppose there are $x_0 \in D$, $v_0 \in E\setminus\{0\}$ such that $\kappa_D(x_0;v_0) = 0$. Let U be a bounded neighborhood of x_0 in E, say $U \subset B(x_0,R)$ for some $R > 0$, where $B(x_0,R) = B_{\|\ \|}(x_0,R)$, and let $r > 0$ be such that $B_k(x_0,r) \subset U$.

For every $\varepsilon > 0$ there exist $f \in \mathrm{Hol}(\Delta,D)$, $\tau \in \mathbb{C}$, such that

$$f(0) = x_0, \quad df(0)\tau = v_0, \quad |\tau| < \varepsilon.$$

For any $0<\delta<1$, we have by the Cauchy inequalities

$$\|df(0)\| \leqslant \frac{1}{\delta} \sup\{\|f(\zeta)\| : \zeta \in \Delta_\delta\}$$

where $\Delta_\delta = \{\zeta \in \mathbb{C}: |\zeta| < \delta\}$. Hence

$$\frac{1}{\delta} \sup\{\|f(\zeta)\| : \zeta \in \Delta_\delta\} \geqslant \frac{\|v_0\|}{|\tau|} > \frac{\|v_0\|}{\varepsilon}$$

i.e.

$$\sup\{\|f(\zeta)\| : \zeta \in \Delta_\delta\} > \frac{\delta}{\varepsilon} \|v_0\|.$$

For any $0<\delta<1$ choose $\varepsilon > 0$ such that

$$\frac{\delta}{\varepsilon} \|v_0\| > R, \quad \text{i.e.} \quad 0 < \varepsilon < \frac{\delta}{R} \|v_0\|.$$

Hence there is some $\zeta_1 \in \Delta_\delta$ for which $\|f(\zeta_1)\| > R$, and therefore $f(\zeta_1) \notin U$. The segment $(0,\zeta_1)$ contains some point ζ_2 such that $f(\zeta_2) \in \partial B_k(0,r)$. On the other hand

$$\omega(0,\zeta_2) = \frac{1}{2}\log\frac{1+|\zeta_2|}{1-|\zeta_2|} < \frac{1}{2}\log\frac{1+\delta}{1-\delta},$$

and therefore

$$r = k_D(0,f(\zeta_2)) \leqslant \omega(0,\zeta_2) < \frac{1}{2}\log\frac{1+\delta}{1-\delta} \to 0$$

as $\delta \to 0$. This contradiction proves the proposition.

Q.E.D.

Proposition V.1.10. *Let* D *be a hyperbolic domain in a complex normed space* E, *and let* $g \in \mathrm{Hol}(\Delta \times D, D)$ *be such that for some* $\zeta_0 \in \Delta$, $g(\zeta_0,\cdot) \in \mathrm{Aut}(D)$. *Then* g *is independent of* ζ.

Proof. Consider the maps $h \in \mathrm{Hol}(\Delta\times D, \Delta\times D)$, $h_0 \in \mathrm{Aut}(\Delta\times D)$ defined by

$$h(\zeta,x) = (\zeta,g(\zeta,x)), \quad h_0(\zeta,x) = (\zeta,g(\zeta_0,x)).$$

The map $h\circ h_0^{-1} \in \mathrm{Hol}(\Delta\times D, \Delta\times D)$ is the identity on $\{\zeta_0\}\times D$. For any $x_0 \in D$, the power series expansion of $h\circ h_0^{-1}$ in a neighborhood of (ζ_0,x_0) is

$$\zeta \mapsto \zeta, \quad x \mapsto x + (\zeta-\zeta_0)(u + \sum_{q=1}^{+\infty} P_q(\zeta-\zeta_0, x-x_0)),$$

where $u \in E$, and $P_q \in P^q(\mathbb{C} \times E, E)$. By Proposition V.1.2, for all $\tau \in \mathbb{C}$, $v \in E$,

$$\kappa_{\Delta\times D}((\zeta_0,x_0);(\tau,v)) \geqslant \kappa_{\Delta\times D}((\zeta_0,x_0);(d(h\circ h_0^{-1})(\zeta_0,x_0))(\tau,v)),$$

i.e., by Proposition V.1.6,

$$\max\left(\frac{|\tau|}{1-|\zeta_0|^2}, \kappa_D(x_0;v)\right) \geqslant \max\left(\frac{|\tau|}{1-|\zeta_0|^2}, \kappa_D(x_0; v+\tau u)\right).$$

Choosing $v = \tau^2 u$, we have for all $\tau \neq 0$

$$\max\left(\frac{1}{1-|\zeta_0|^2}, |\tau|\kappa_D(x_0;u)\right) \geqslant \max\left(\frac{1}{1-|\zeta_0|^2}, |\tau+1|\kappa_D(x_0;u)\right).$$

For τ (real and) $\gg 0$, that implies

$$\kappa_D(x_0;u) = 0,$$

and therefore, by Proposition V.1.9, $u = 0$. Hence

$$d(h\circ h_0^{-1})(\zeta_0,x_0) = \mathrm{Id},$$

and by Theorem IV.2.4, $h\circ h_0^{-1}$ = Identity, i.e., $h = h_0$. Thus

$$g(\zeta,x) = g(\zeta_0,x) \qquad \text{for all } x \in D \text{ and all } \zeta \in \Delta.$$

Q.E.D.

§ 2. Local properties.

Lemma V.2.1. *For every* $x \in D$, $\gamma_D(x;\cdot)$ *is a continuous semi-*

norm on E. *If* D *is bounded,* $\gamma_D(x;\cdot)$ *is an equivalent norm to* $\| \ \|$.

Proof. The first part of the lemma is a consequence of (V.1.4), (V.1.5) and (V.1.9). If D is bounded, for any $x \in D$ there is some $R > 0$ such that $D \subset B(x,R)$. Thus

$$\gamma_D(x;v) \geqslant \gamma_{B(x,R)}(x;v) = \frac{\|v\|}{R} \qquad \text{for all } v \in E.$$

Q.E.D.

Let $x_0 \in D$ and let $v \in E\setminus\{0\}$. If $r > 0$ is such that $\overline{B(x_0,r)} \subset D$, then, for any $x \in B(x_0,\frac{r}{2})$, $x+\frac{r}{2\|v\|}\zeta v \in B(x_0,r)$ for all $\zeta \in \overline{\Delta}$. Let $f \in \mathrm{Hol}(D,\mathbb{C})$, and choose $r > 0$ so small that f is bounded on $\overline{B(x_0,r)}$. Then for $x \in B(x_0,\frac{r}{2})$, $q = 1,2,\ldots$,

$$\hat{d}^q f(x)(v) = \left(\frac{2\|v\|}{r}\right)^q \hat{d}^q f(x)\left(\frac{r}{2\|v\|}v\right) =$$

$$= \frac{q!}{2\pi}\left(\frac{2\|v\|}{r}\right)^q \int_0^{2\pi} e^{-iq\theta} f\left(x+\frac{r}{2\|v\|}e^{i\theta}v\right)d\theta,$$

and therefore

$$\text{(V.2.1)} \qquad |\hat{d}^q f(x)(v)| \leqslant \left(\frac{2\|v\|}{r}\right)^q q! \sup\{\|f(y)\| : y \in B(x_0,r)\}$$

for $q=1,2,\ldots$, $x \in B(x_0,\frac{r}{2})$, $v \in E$.

Now, let $f \in \mathrm{Hol}(D,\Delta)$, and consider the complex-valued holomorphic function $x \mapsto \hat{d}^q f(x)(v)$. For $x_1,x_2 \in B(x_0,\frac{r}{4})$, the function $t \mapsto \hat{d}^q f(x_1+t(x_2-x_1))(v)$ is C^∞ on $[0,1]$.

Thus, by the mean value theorem

$$\hat{d}^q f(x_2)(v) - \hat{d}^q f(x_1)(v) = d(\hat{d}^q f(x)(v))(x_2-x_1)$$

for some $x = x_1+t(x_2-x_1)$ with $0<t<1$. Since $\|x-x_0\| < \frac{r}{4}$, then, by (V.2.1)

$$|d(\hat{d}^q f(x)(v))(x_2 - x_1)| \leqslant q! \frac{2\|x_2 - x_1\|}{r} \left(\frac{2\|v\|}{r}\right)^q ,$$

and therefore

$$\text{(V.2.2)} \qquad |\hat{d}^q f(x_2)(v) - \hat{d}^q f(x_1)(v)| \leqslant \left(\frac{2}{r}\right)^{q+1} q! \|x_2 - x_1\| \|v\|^q$$

for all $f \in \mathrm{Hol}(D,\Delta)$, all $v \in E$, all $x_1, x_2 \in B(x_0, \frac{r}{4})$, $q = 1, 2, \ldots$. Note that r does not depend on f.

Proposition V.2.2. *Let* D *be a domain in* E. *The function* $\gamma_D : D \times E \to \mathbb{R}_+$ *is locally Lipschitz.*

Proof. 1) Let $v \in E$, and let $x_0 \in D$ and $r > 0$ be such that $B(x_0, r) \subset D$. Let $x_1, x_2 \in B(x_0, \frac{r}{4})$ and suppose that $\gamma_D(x_2; v) \geqslant \gamma_D(x_1; v)$. Since $\mathrm{Aut}(\Delta)$ is transitive on Δ, then, by (V.2.2)

$$\begin{aligned}
\gamma_D(x_2;v) - \gamma_D(x_1;v) &= \sup\{|df(x_2)(v)| : f \in \mathrm{Hol}(D,\Delta),\ f(x_2) = 0\} - \\
&- \sup\left\{\frac{|df(x_1)(v)|}{1-|f(x_1)|^2} : f \in \mathrm{Hol}(D,\Delta),\ f(x_2) = 0\right\} \leqslant \\
&\leqslant \sup\left\{|df(x_2)(v)| - \frac{|df(x_1)(v)|}{1-|f(x_1)|^2} : f \in \mathrm{Hol}(D,\Delta),\ f(x_2) = 0\right\} \leqslant \\
&\leqslant \sup\{|df(x_2)(v)| - |df(x_1)(v)| : f \in \mathrm{Hol}(D,\Delta),\ f(x_2) = 0\} \leqslant \\
&\leqslant \sup\{|df(x_2)(v) - df(x_1)(v)| : f \in \mathrm{Hol}(D,\Delta), f(x_2) = 0\} \leqslant \\
&\leqslant \left(\frac{2}{r}\right)^2 \|x_2 - x_1\| \|v\| .
\end{aligned}$$

Thus

$$|\gamma_D(x;v) - \gamma_D(x_0;v)| \leqslant \left(\frac{2}{r}\right)^2 \|x - x_0\| \|v\|$$

for all $x \in B(x_0, \frac{r}{4})$.

2) For $v_0, v_1 \in E$ we have, by (V.1.5) and (V.2.1),

$$\begin{aligned}
&|\gamma_D(x_0;v_0) - \gamma_D(x_0;v_1)| \leqslant \gamma_D(x_0;v_0 - v_1) = \\
&= \sup\{|df(x_0)(v_0 - v_1)| : f \in \mathrm{Hol}(D,\Delta),\ f(x_0) = 0\} \leqslant \frac{2}{r} \|v_0 - v_1\| .
\end{aligned}$$

3) For $x_1 \in B(x_0, \frac{r}{4})$, $v_0, v_1 \in E$, we have, by 1) and 2),

$$|\gamma_D(x_0;v_0) - \gamma_D(x_1;v_1)| \leq |\gamma_D(x_0;v_0) - \gamma_D(x_0;v_1)| +$$

$$+ |\gamma_D(x_0;v_1) - \gamma_D(x_1;v_1)| \leq \frac{2}{r}\|v_1 - v_0\| + (\frac{2}{r})^2\|x_1 - x_0\|\|v_1\|.$$

Q.E.D.

We prove now

Theorem V.2.3. *The function* $\log\gamma_D : D \times E \to \mathbb{R}$ *is plurisubharmonic.*

Proof. First of all, γ_Δ being given by (V.1.7), a direct computation proves immediately the theorem in the case $D = \Delta$.

Now, for $u,y \in E$, the set $U_{u,y} = \{\zeta \in \mathbb{C}: u+\zeta y \in D\}$ is open in $\mathbb{C}$. For any $f \in \mathrm{Hol}(D,\Delta)$ the function

$$x \mapsto \log \gamma_\Delta(f(x),1) = \log\frac{1}{1-|f(x)|^2}$$

is plurisubharmonic on D. Furthermore for any $v \in E$ the complex valued function $(x,v) \mapsto df(x)v$ is holomorphic on $D \times E$. By Lemma II.6.1, the function $(x,v) \mapsto \log|df(x)v|$ is plurisubharmonic on $D \times E$. Hence the function $(x,v) \mapsto$ $\mapsto \log\frac{|df(x)v|}{1-|f(x)|^2} = \log\gamma_\Delta(f(x);df(x)v)$ is plurisubharmonic. Thus the continuous function $\log\gamma_\Delta$, being the upper envelope of a family of plurisubharmonic functions on $D \times E$, is plurisubharmonic.

Q.E.D.

We shall now prove that the function $\kappa_D: D \times E \to R_+$ is upper semi-continuous. This fact was established by H.L.Royden in [Royden, 1] in the finite-dimensional case. Our proof will follow, with only minor changes, Royden's argument.

Lemma V.2.4. *Let* $R > 0$ *and let* $h \in \mathrm{Hol}(\Delta_R, D)$ ($\Delta_R =$ $= \{\zeta \in \mathbb{C}: |\zeta| < R\}$) *be such that* $h'(0) \neq 0$. *There exist a continuous linear form* $\lambda \neq 0$ *on* E *and, for every* $0<r<R$, *a neighborhood* U *of* 0 *in* $F = \operatorname{Ker}\lambda$ *and a function* $g \in \operatorname{Hol}(\Delta_r \times U, D)$, *such that:* $g|_{\Delta_r \times \{0\}} = h|_{\Delta_r}$; *the differential* $dg(0,0)$ *of* g *at* $(0,0)$ *is a bi-continuous isomorphism of* $\mathbb{C} \times F$ *onto* E, *and* g *maps a neighborhood of* $(0,0)$ *biholomorphically onto a neighborhood of* $h(0)$.

Proof. Assume $h(0) = 0$. Let λ be a continuous linear form on E such that $\lambda(h'(0)) = 1$, and let $f \in \operatorname{Hol}(\Delta_R \times F, E)$ be defined by

$$f(\zeta,y) = h(\zeta) + y \qquad (\zeta \in \Delta_R\ ,\ y \in F).$$

Let $h(\zeta) = \sum_{\nu=1}^{+\infty} \zeta^\nu a_\nu$ ($a_\nu \in E$ for $\nu=1,2,\dots$; $a_1 \neq 0$) be the power series expansion of h in Δ_R. The differential $df(0,0)$ of f at $(0,0)$ is defined by

$$df(0,0)(\zeta,y) = \zeta a_1 + y \qquad (\zeta \in \mathbb{C},\ \ y \in F).$$

Hence $df(0,0)$ is a bi-continuous isomorphism of $\mathbb{C} \times F$ onto E, and

$$df(0,0)^{-1}(x) = (\lambda(x),\ x-\lambda(x)a_1) \qquad (x \in E).$$

For $x = f(\zeta,y)$

$$df(0,0)^{-1}(x) = (\zeta,y) + \sum_{\nu=2}^{+\infty} \zeta^\nu df(0,0)^{-1}(a_\nu),$$

i.e.

$$\text{(V.2.3)} \qquad \lambda(x) = \sigma(\zeta),\quad x-\lambda(x)a_1 = y + h(\zeta) - \sigma(\zeta)a_1,$$

where $\sigma \in \operatorname{Hol}(\Delta_R\ ,\ \mathbb{C})$ is expressed by

$$\sigma(\zeta) = \zeta + \sum_{\nu=2}^{+\infty} \zeta^\nu \lambda(a_\nu).$$

Being $\sigma'(0) = 1$, there is an open neighborhood A of 0 in $\mathbb{C}$ which is mapped by σ bi-holomorphically onto an open neighborhood A_1 of 0 in Δ_R. Let $\tau = (\sigma_{|A})^{-1} : A_1 \to A$, and let V be a neighborhood of 0 in E such that $\lambda(V) \subset A_1$. Then for $x \in V$ (V.2.3) yields

$$\zeta = \tau\circ\lambda(x),$$

$$y = x - \lambda(x)a - h\circ\tau\circ\lambda(x) + \sigma\circ\tau\circ\lambda(x)a_1 =$$

$$= x - h(\tau(\lambda(x))).$$

That proves that there is a neighborhood W of 0 in $\Delta_R \times F$ whose image $f(W)$ is a neighborhood of 0 in E, and the restriction $f_{|W}$ is a bi-holomorphic map of W onto $f(W)$.

Since $f(\Delta_R \times \{0\}) = h(\Delta_R) \subset D$, then, for every $0<r<R$, D contains the compact set $f(\overline{\Delta_r} \times \{0\})$. Hence there is an open neighborhood U of 0 in F such that $\overline{\Delta_r} \times U \subset f^{-1}(D)$, i.e.

$$f(\overline{\Delta_r} \times U) \subset D. \quad \text{Take} \quad g = f_{|\overline{\Delta_r} \times U}$$

Q.E.D.

As a neighborhood U satisfying Lemma V.2.4, we can choose

$$U = B_{F,s} = \{y \in F : \|y\| < s\}$$

for a suitable $s > 0$.

Lemma V.2.5. *Given* $\varepsilon > 0$, $x \in D$, $v \in E$, *there exist:* $r > 0$, $s > 0$, *a continuous linear form* λ *on* E $(\lambda \neq 0)$, *and* $g \in \mathrm{Hol}(\Delta_r \times B_{F,s}, D)$ *such that:*

$$g(0,0) = x; \quad dg(0,0)(1,0) = v;$$

g *maps an open neighborhood of* $(0,0)$ *bi-holomorphically onto an open neighborhood of* x;

$$\kappa_{\Delta_r \times B_{F,s}}((0,0);(1,0)) < \kappa_D(x;v) + \varepsilon .$$

<u>Proof</u>. Since $\kappa_{\Delta_R}(0;1) = \frac{1}{R}$, then

$$\kappa_D(x;v) = \inf\{\frac{1}{R}: h \in \mathrm{Hol}(\Delta_R, D), h(0)=x, h'(0)=v\}.$$

Choose now h such that

$$\frac{1}{R} < \kappa_D(x;v) + \varepsilon,$$

and let r be such that $0<r<R$ and

$$\frac{1}{r} < \kappa_D(x;v) + \varepsilon.$$

Starting from h we define g as in Lemma V.2.4. Then, by Proposition V.1.6,

$$\kappa_{\Delta_r \times B_{F,s}}((0,0);(1,0)) = \kappa_{\Delta_r}(0;1) = \frac{1}{r} < \kappa_D(x;v) + \varepsilon.$$

Q.E.D.

<u>Proposition V.2.6</u>. *The function* $\kappa_D\colon D \times E \to \mathbb{R}_+$ *is upper semi-continuous.*

<u>Proof</u>. (Same notations as in the proof of Lemma V.2.5). Since $\kappa_{\Delta_r \times B_{F,s}}$ is continuous at $(0,0,0,0)$, for any $\varepsilon > 0$ there is an open neighborhood W of $(0,0,0,0)$ in $\Delta_r \times B_{F,s} \times \mathbb{C} \times F$ such that for $(\zeta,y,\tau,w) \in W$,

$$\kappa_{\Delta_r \times B_{F,s}}((\zeta,y);(\tau,w)) < \kappa_{\Delta_r \times B_{F,s}}((0,0);(1,0)) + \varepsilon.$$

Since g is bi-holomorphic on a neighborhood of $(0,0)$ in $\Delta_r \times F$, then - if W is sufficiently small - (g,dg) maps W onto a set containing a neighborhood V of (x,v) in $D \times E$.

Let $(x',v') \in V$. Then $x' = g(\zeta,y)$, $v' = dg(\zeta,y)(\tau,w)$ for some $(\zeta,y,\tau,w) \in W$, and therefore

$$\kappa_D(x';v') = \kappa_D(g(\zeta,y);dg(\zeta,y)(\tau,w)) \leqslant$$
$$\leqslant \kappa_{\Delta_r \times B_{F,s}}((\zeta,y);(\tau,w)) <$$

$$< \kappa_{\Delta_r \times B_{F,s}}((0,0);(1,0)) + \varepsilon < \kappa_D(x;v) + 2\varepsilon.$$

Q.E.D.

§ 3. Inner distances.

Let X be a metric space with distance d. Before investigating the relationship between Kobayashi metrics and Kobayashi pseudo-distances we review the basic properties of inner distances in metric spaces. A thorough account of the theory can be found in [Rinow, 1], cf. also [Kobayashi, 4]. Given a continuous curve $\ell: [a,b] \to X$, the length $L(\ell)$ is, by definition

$$L(\ell) = \sup \sum_{j=1}^{n} d(\ell(t_{j-1}), \ell(t_j)),$$

over all partitions $a = t_0 < t_1 < \dots < t_n = b$ of $[a,b]$; ℓ is said to be *rectifiable* if $L(\ell) < \infty$.

Example. Let $\ell: [a,b] \to \mathbb{R}$ be a continuous curve on the real line. Then ℓ is rectifiable if, and only if, ℓ is of bounded variation over every compact interval contained in $[a,b]$.

Exercise. For any $x \in X$ the function $t \mapsto d(x,\ell(t))$ is of bounded variation on every compact interval contained in $[a,b]$.

The metric space X is called *finitely arcwise connected* if, for every pair $x,y \in X$, there is a rectifiable curve from x to y. Let $B_d(x,r) = \{y \in X: d(x,y) < r\}$. The metric space X is said to be *without detour* if, for every $x \in X$ and every $\varepsilon > 0$, there is $\delta > 0$ such that every $y \in B_d(x,\delta)$ can be joined to x by a rectifiable curve ℓ with $L(\ell) < \varepsilon$.

Lemma V.3.1. *If X is without detour then X is locally connected.*

Proof. For $x \in X$ and $\varepsilon > 0$, let $\delta > 0$ be chosen as before. Let S be the union of the images of $[0,1]$ under all rectifiable curves $\ell : [0,1] \to X$, with $\ell(0) = x$ and $L(\ell) < \varepsilon$. Then for $0 \leqslant t \leqslant 1$,

$$d(x, \ell(t)) \leqslant L(\ell_{|[0,t]}) \leqslant L(\ell) < \varepsilon,$$

and therefore

$$S \subset B_d(x, \varepsilon).$$

Furthermore $B_d(x, \delta) \subset S$; hence S is a neighborhood of x. Since $\ell([0,1])$ is connected for every continuous curve ℓ, then S is connected.

Q.E.D.

Lemma V.3.2. *If* X *is connected and without detour, then* X *is finitely arcwise connected.*

Proof. For $x \in X$, let $X(x)$ be the set of points in X which can be joined to x by a rectifiable curve. Since X is without detour, then $X(x)$ is open. If $y \notin X(x)$, y has a neighborhood disjoint from $X(x)$, since X is without detour. Thus $X(x)$ is open and closed. Thus, being $x \in X(x)$, and therefore $X(x) \neq \emptyset$, $X(x) = X$, for X is connected.

Q.E.D.

Let X be finitely arcwise connected. For $x, y \in X$, let

$$d^i(x,y) = \inf L(\ell)$$

over all rectifiable curves joining x and y.

Lemma V.3.3. *If* X *is finitely arcwise connected,* d^i *is a distance on* X*, and*

$$d(x,y) \leqslant d^i(x,y). \tag{V.3.1}$$

Proof. Clearly $d^i(x,y) = d^i(y,x) \geqslant 0$, and $d^i(x,x) = 0$. The last assertion of the lemma follows trivially from the triangle inequality (for d). Hence $d^i(x,y) = 0$ implies $d(x,y) = 0$, and therefore $x = y$. Given $x,y,z \in X$ and $\varepsilon > 0$, there exist rectifiable curves ℓ^1, ℓ^2 joining x and y, y and z, such that

$$L(\ell^1) < d^i(x,y) + \varepsilon, \quad L(\ell^2) < d^i(y,z) + \varepsilon.$$

The curves ℓ^1 and ℓ^2 define a rectifiable curve ℓ joining x and z, for which

$$d^i(x,z) \leqslant L(\ell) \leqslant L(\ell^1) + L(\ell^2) < d^i(x,y) + d^i(y,z) + 2\varepsilon,$$

for all $\varepsilon > 0$. That proves that

$$d^i(x,z) \leqslant d^i(x,y) + d^i(y,z)$$

and completes the proof of the lemma.

The distance d^i is called the *inner distance* defined by d.

Clearly $B_{d^i}(x,r) \subset B_d(x,r)$ for all $x \in X$ and all $r>0$. The metric space X is without detour if, and only if, for any $x \in X$ and any $\varepsilon > 0$ there is some $\delta > 0$ such that

$$B_d(x,\delta) \subset B_{d^i}(x,\varepsilon).$$

In conclusion we have

Lemma V.3.4. *Let* X *be a finitely arcwise connected metric space. Then the* d^i*-topology is finer than the* d*-topology. The two topologies are equivalent if, and only if,* X *is without detour.*

Let $L^i(\ell)$ be the length of a curve ℓ with respect to d^i. By (V.3.1) $L(\ell) \leqslant L^i(\ell)$. On the other hand, let $\ell: [0,1] \to X$ be a continuous parametrization of ℓ. For

$0 \leqslant a < b \leqslant 1$, we have

$$d^i(\ell(a), \ell(b)) \leqslant L(\ell_{|[a,b]}),$$

where $\ell_{|[a,b]}$ denotes the restriction of ℓ to $[a,b]$. Hence $L^i(\ell) \leqslant L(\ell)$. In conclusion we have proved

Lemma V.3.5. *If X is finitely arcwise connected, for every continuous curve ℓ in X,*

$$L^i(\ell) = L(\ell).$$

Inequality (V.3.1) and Lemma V.3.4 yield

Lemma V.3.6. *Let X be finitely arcwise connected without detour. If X is complete for d, then it is complete for d^i. If X is finitely compact for d, then it is finitely compact for d^i.*

Recall that X is said to be *finitely compact* if every bounded infinite set has some accumulation point.

Definition. The distance d is said to be *inner* if $d = d^i$.

By Lemma V.3.4, if d is inner, then X is without detour.

Lemma V.3.5 implies

Lemma V.3.7. *If X is finitely arcwise connected then d^i is an inner distance.*

Proposition V.3.8. *Let X be a finitely arcwise connected metric space with an inner distance d. Then for all $x \in X$, $r > 0$, $B_d(x,r)$ is finitely arcwise connected.*

Proof. For any $y \in B_d(x,r)$ there is a rectifiable curve ℓ joining x and y in X such that $d(x,y) = d^i(x,y) \leqslant L(\ell) <$ $< r$. For any $z \in \ell$, $d(x,z) < L(\ell) < r$, and therefore $z \in$ $\in B_d(x,r)$, i.e., $\ell \in B_d(x,r)$. Thus every two points in

$B_d(x,z)$ can be joined by a rectifiable curve in $B_d(x,r)$.

Q.E.D.

§ 4. The Kobayashi metric and the Kobayashi distance.

We will now prove that the Kobayashi distance is inner. Let D be a domain in a complex normed space E, on which the Kobayashi pseudo-distance is a distance. Let $\ell: [0,1] \to D$ be a piecewise C^1 function and let $L_\kappa(\ell)$ be defined by

$$L_\kappa(\ell) = \int_0^1 \kappa_D(\ell(t);\ \ell'(t))dt.$$

Note that, by Proposition V.2.6, the integral on the right converges. For $x,y \in D$, we define $\tilde{k}_D(x,y)$ by

$$\tilde{k}_D(x,y) = \inf \int_0^1 \kappa_D(\ell(t);\ell'(t))dt$$

where the infimum is taken over all piecewise C^1 curves from x to y, i.e. over all piecewise C^1 functions $\ell: [0,1] \to D$ such that $\ell(0) = x$, $\ell(1) = y$.

Exercise. *Prove that there is a* C^∞ *curve joining* x *and* y *and* $\tilde{k}_D(x,y)$ *is the infimum over all* C^p *curves from* x *to* y, *for every* p *fixed,* $1 \leqslant p \leqslant \infty$.

By a similar argument to the proof of Lemma V.3.3 one shows that $\tilde{k}_D(x,y)$ is a pseudo-distance. We prove now

Theorem V.4.1. *The Kobayashi pseudo-distance is the integrated form of* κ_D , i.e.,

$$k_D(x,y) = \tilde{k}_D(x,y).$$

This theorem has been proved by H.L. Royden for finite-dimensional complex manifolds; cf. Theorem 1 of [Royden, 1].

Royden's proof rests essentially on the key Lemma 1 of [Royden, 1] (which was proved in [Royden, 1] for domains in $\mathbb{C}^n$ and in [Royden, 2] for finite dimensional manifolds). Our Lemma V.2.5 extends Lemma 1 of [Royden, 1] to domains in E. Substituting Lemma V.2.5 for Lemma 1 of [Royden, 1], Royden's proof of Theorem 1 can be adapted to establish Theorem V.4.1. We give the proof here for the sake of completeness.

Proof. I. We prove first that $\tilde{k}_D(x,y) \leqslant k_D(x,y)$. Let $\varepsilon > 0$ and $\{\zeta_1', \zeta_1'', \dots, \zeta_\nu', \zeta_\nu'', f_1, \dots, f_\nu\}$ be an analytical chain joining x and y in D (notations as in § 1 of chapter IV) such that

$$\sum_{j=1}^{\nu} \omega(\zeta_j', \zeta_j'') < k_D(x,y) + \varepsilon .$$

Let ℓ be the geodesic for the Poincaré metric, joining ζ_j', ζ_j'' in Δ, and let ℓ be the piecewise differentiable curve defined by $f_j|\ell_j$ $(j = 1, \dots, \nu)$. Then

$$\tilde{k}_D(x,y) \leqslant \int_\ell \kappa_D = \sum_{j=1}^{\nu} \int_{f_j(\ell_j)} \kappa_D \leqslant \sum_{j=1}^{\nu} \int_{\ell_j} \kappa_\Delta =$$

$$= \sum_{j=1}^{\nu} \omega(\zeta_j', \zeta_j'') < k_D(x,y) + \varepsilon .$$

Thus $\tilde{k}_D(x,y) \leqslant k_D(x,y)$, since $\varepsilon > 0$ is arbitrary.

II. We prove now that $\tilde{k}_D(x,y) \geqslant k_D(x,y)$. For any $\varepsilon > 0$ there is a differentiable function $\varphi\colon [0,1] \to D$ such that $\varphi(0) = x$, $\varphi(1) = 1$, and

$$\int_0^1 \kappa_D(\varphi(t);\varphi'(t))dt < \tilde{k}_D(x,y) + \varepsilon .$$

The function $t \mapsto \kappa_D(\varphi(t);\varphi'(t))$, being upper semi-continuous on $[0,1]$, is the pointwise limit of a decreasing sequence of continuous functions. Hence, by the monotone convergence theorem, there is a positive continuous function σ on $[0,1]$ such that $\sigma(t) > \kappa_D(\varphi(t);\varphi'(t))$ for all $t \in [0,1]$, and

$$\int_0^1 \sigma(t)dt < \tilde{k}_D(x,y) + \varepsilon.$$

The function σ, being continuous, is Riemann integrable. Hence there is a $\delta > 0$ such that for any choice of $0 = t_0 < t_1 < \dots < t_n = 1$ with $t_{j+1}-t_j < \delta$ and any choice of $s_j \in$ $\in (t_{j-1},t_j)$, then

$$\text{(V.4.1)} \qquad \sum_{j=1}^{n} (t_j-t_{j-1})\sigma(s_j) < \tilde{k}_D(x,y) + \varepsilon.$$

III. We prove now that for any $s \in [0,1]$ there is an open interval J_s such that $s \in J_s$, and for all $t',t'' \in J_s \cap [0,1]$

$$k_D(\varphi(t'),\varphi(t'')) < (1 + \varepsilon)|t' - t''|\sigma(s).$$

By Lemma V.2.5 there exist: $r > 0$, $r' > 0$, a continuous linear form $\lambda \neq 0$ on E and a function $g \in \mathrm{Hol}(\Delta_r \times B_{F,r'},D)$, where $F = \mathrm{Ker}\,\lambda$, such that:

$$g(0) = \varphi(s), \quad dg(0,0)(1,0) = \varphi'(s);$$

g is a bi-holomorphic homeomorphism of a neighborhood of $(0,0)$ in $\Delta_r \times B_{F,r'}$ onto a neighborhood of $\varphi(s)$ in D such that

$$\frac{1}{r} = \kappa_{\Delta_r \times B_{F,r'}}((0,0);(1,0)) < \kappa_D(\varphi(s);\varphi'(s)) + \sigma(s) -$$
$$- \kappa_D(\varphi(s);\varphi'(s)) = \sigma(s).$$

Hence there is an open interval $I_s \ni s$ and a differentiable map $\mu: I_s \to \Delta_r \times B_{F,r'}$ such that $\varphi = g \circ \mu$ on $I_s \cap [0,1]$, and

$$\mu(s) = (0,0), \quad \mu'(s) = (1,0) \in \mathbb{C} \times F;$$

i.e., there are differentiable maps $\mu_1: I_s \to \Delta_r$,

$\mu_2 : I_{r'} \to B_{F,r'}$ such that $\mu(t) = (\mu_1(t), \mu_2(t))$ and $\mu_1(t) = t-s+O(|t-s|^2)$, $\|\mu_2(t)\| = O(|t-s|^2)$.

For $f \in \mathrm{Aut}(\Delta_r)$, the map $(\zeta, x) \mapsto (f(\zeta), x)$ defines an automorphism of $\Delta_r \times B_{F,r'}$. Let $t', t'' \in I_s \cap [0,1]$. Choosing $f \in \mathrm{Aut}(\Delta_r)$ in such a way that $f(\mu_1(t')) = 0$, we have

$$k_{\Delta_r \times B_{F,r'}}(\mu(t'), \mu(t'')) = k_{\Delta_r \times B_{F,r'}}((0, \mu_2(t')), (f(\mu_1(t''), \mu_2(t''))) \leqslant$$

$$\leqslant k_{\Delta_r \times B_{F,r'}}((0,\mu_2(t')), (0,\mu_2(t''))) + k_{\Delta_r \times B_{F,r'}}((0,\mu_2(t'')), (f(\mu_1(t'')), \mu_2(t''))) \leqslant$$

$$\leqslant k_{B_{F,r'}}(\mu_2(t'), \mu_2(t'')) + k_{\Delta_r}(0, f(\mu_1(t''))) =$$

$$= k_{B_{F,r'}}(\mu_2(t'), \mu_2(t'')) + k_{\Delta_r}(\mu_1(t'), \mu_1(t'')))$$

where

$$k_{\Delta_r}(\mu_1(t'), \mu_2(t'')) = \omega\left(\frac{\mu_1(t')}{r}, \frac{\mu_1(t'')}{r}\right),$$

$$k_{B_{F,r'}}(\mu_2(t'), \mu_2(t''))) \leqslant k_{B_{F,r'}}(0, \mu_2(t')) + k_{B_{F,r'}}(0, \mu_2(t'')) \leqslant$$

$$\leqslant \omega\left(0, \frac{\mu_2(t')}{r'}\right) + \omega\left(0, \frac{\mu_2(t'')}{r'}\right).$$

Hence there is an open interval J_s, with $s \in J_s \subset I_s$ such that for $t', t'' \in J_s \cap [0,1]$

$$k_{\Delta_r \times B_{F,r'}}(\mu(t'), \mu(t'')) \leqslant (1+\varepsilon) \frac{|t'-t''|}{r} < (1+\varepsilon)|t'-t''|\sigma(s),$$

and therefore

$$k_D(\varphi(t'), \varphi(t'')) = k_D(g(\mu(t')), g(\mu(t''))) \leqslant$$

$$\leqslant k_{\Delta_r \times B_{F,r'}}(\mu(t'), \mu(t'')) < (1+\varepsilon)|t'-t''|\sigma(s).$$

IV. By the Lebesgue covering lemma, there is $\rho > 0$ such that whenever $|t'-t''| < \rho$, $t', t'' \in [0,1]$ there is some $s \in [0,1]$ for which $t', t'' \in J_s$. Choosing in II. $\delta < \rho$, then by III. and by (V.4.1):

$$k_D(x,y) = k_D(\varphi(0),\varphi(1)) \leqslant \sum_{j=1}^{n} k_D(\varphi(t_{j-1}),\varphi(t_j)) \leqslant$$
$$\leqslant (1+\varepsilon) \sum_{j=1}^{n} (t_j - t_{j-1})\sigma(s_j) < (1+\varepsilon)(\tilde{k}_D(x,y)+\varepsilon).$$

Since $\varepsilon > 0$ is arbitrary, that completes the proof of the theorem.

Corollary V.4.2. *If* D *is hyperbolic the Kobayashi distance is inner.*

As a consequence of Theorem V.3.1 and Proposition V.1.6, we have also

Proposition V.4.3. *Let* D_1 *and* D_2 *be two domains in two complex normed spaces* E_1 *and* E_2. *Then for all* $x_j, y_j \in D_j$ $(j=1,2)$ *we have*

$$k_{D_1 \times D_2}((x_1,x_2),(y_1,y_2)) = \max(k_{D_1}(x_1,y_1), k_{D_2}(x_2,y_2)).$$

Let us consider now the Carathéodory differential metric γ_D, and let us define $\tilde{c}_D$ on $D \times D$ by

$$\tilde{c}_D(x,y) = \inf \int_0^1 \gamma_D(\ell(t);\ell'(t))dt,$$

where the infimum is taken over all differentiable curves $\ell: [0,1] \to D$ joining x and y in D, or, equivalently, over all piecewise differentiable curves $\ell: [0,1] \to D$ joining x and y in D. First of all, we have obviously $\tilde{c}_\Delta = c_\Delta = \omega$. For any $f \in \mathrm{Hol}(D,\Delta)$, $f \circ \ell$ is a differentiable (or, respectively, piecewise differentiable) curve joining $f(x)$ and $f(y)$ in Δ. Since $\gamma_\Delta(f(\ell(t));(f(\ell(t)))') = \gamma_\Delta(f(\ell(t));f'(\ell(t))\ell'(t)) \leqslant$ $\leqslant \gamma_D(\ell(t);\ell'(t))$, then

$$c_\Delta(f(x),f(y)) \leqslant \int_0^1 \gamma_\Delta(f(\ell(t));(f(\ell(t)))')dt \leqslant \int_0^1 \gamma_D(\ell(t);\ell'(t))dt$$

for all differentiable (or piecewise differentiable) curves joining x and y in D. Hence

$$c_\Delta(f(x),f(y)) \leqslant \tilde{c}_D(x,y)$$

for all $f \in \mathrm{Hol}(D,\Delta)$, and therefore

$$\tilde{c}_D(x,y) \geqslant c_D(x,y) \quad \text{for all} \quad x,y \in D. \tag{V.4.2}$$

By (V.1.3) and Theorem V.4.1 we have also

$$\tilde{c}_D(x,y) \leqslant k_D(x,y) \quad \text{for all} \quad x,y \in D.$$

Example. We consider now an example of a bounded domain D in a Banach space E, such that the Carathéodory distance c_D on D is not inner (and so, in particular, $c_D \neq \tilde{c}_D$). It is obviously sufficient to consider a bounded domain D on which there exist $x \in D$, $r > 0$, $r' > 0$ such that

$$B_c(B_c(x,r),r') \neq B_c(x,r+r'),$$

where the notations are as in Lemma IV.1.12, but with reference to the Carathéodory distance rather than to the Kobayashi distance. Then, we can consider the example in the Remark following Lemma IV.1.12.

§ 5. Application. A fixed point theorem.

In this section D will be a bounded domain in a complex Banach space E.

Lemma V.5.1. *Let* $f \in \mathrm{Hol}(D,E)$. *If* $\overline{f(D)} \subset\subset D$ *for every* $x \in D$ $(f^n(x))$ *is a Cauchy sequence for the Kobayashi distance.*

Proof. Let $M > 0$ be such that $\|x\| < M$ for all $x \in D$. Let $\varepsilon > 0$ be such that $\|f(x)-z\| > \varepsilon$ for all $x \in D$, $z \in E\setminus D$, and let $t = \frac{\varepsilon}{2M}$. For any fixed $x \in D$, let $g \in \mathrm{Hol}(D,E)$ be defined on $g \in D$ by

$$g(y) = (1+t)f(y) - tf(x) = f(y) + t(f(y) - f(x)).$$

For any $y \in D$

$$t\|f(y)-f(x)\| \leqslant t(\|f(y)\| + \|f(x)\|) < 2tM = \varepsilon.$$

Thus $g(y) \in D$, i.e., $g(D) \subset D$. Hence, by Proposition V.1.2,

$$\kappa_D(g(x);dg(x)v) \leqslant \kappa_D(x;v) \qquad \text{for all} \quad v \in E.$$

Since $g(x) = f(x)$, $dg(x) = (1+t)df(x)$, then, by (V.1.4),

$$\text{(V.5.1)} \qquad \kappa_D(f(x);df(x)v) \leqslant \frac{1}{1+t}\kappa_D(x;v) \quad \text{for all} \quad x \in D,\ v \in E,$$

and, by Theorem V.4.1,

$$\text{(V.5.2)} \qquad k_D(f(x_1),f(x_2)) \leqslant \frac{1}{1+t}k_D(x_1,x_2) \quad \text{for all} \quad x_1,x_2 \in D.$$

Thus for any $x \in D$ and all $n=1,2,\ldots$

$$k_D(f^n(x),f^{n+1}(x)) \leqslant \frac{1}{(1+t)^n}k_D(x,f(x)).$$

That proves that $(f^n(x))$ $(n=1,2,\ldots)$ is a Cauchy sequence for k_D.

Q.E.D.

Since $f^n(x) \in \overline{f(D)} \subset\subset D$, by Theorem IV.2.2 $(f^n(x))$ is a Cauchy sequence for the norm. Thus $(f^n(x))$ converges in norm to some point $x_0 \in \overline{f(D)} \subset\subset D$. Clearly $f(x_0) = x_0$, and, by (V.5.2) x_0 is the unique fixed point of f in D. That proves the following theorem, due to C.J. Earle and R.S. Hamilton [1]

Theorem V.5.2. *Let* D *be a bounded domain in a complex Banach space* E. *Every* $f \in \mathrm{Hol}(D,E)$ *for which* $\overline{f(D)} \subset\subset D$ *has a unique fixed point in* D.

Remark. A similar inequality to (V.5.1) holds for the Carathéodory differential metric. Hence Lemma V.5.1 holds also with $\tilde{c}_D$ substituting for the Kobayashi distance. Therefore, by (V.4.2) Lemma V.5.1 holds also for the Carathéodory distance. Hence Theorem V.5.2 can be established also by means of the Carathéodory distance. For an extension to locally convex spaces, see [Vesentini, 4].

§ 6. Bounded domains in finitely dimensional vector space.

Let D be a bounded domain in $\mathbb{C}^n$. Let $L^2(D)$ be the Hilbert space of square summable complex valued functions which are square summable on D for the Lebesgue measure, $d\omega$, with the usual scalar product

$$(f_1 \mid f_2) = \int_D f_1(x)\overline{f_2(x)}d\omega(x)$$

and norm $\|f\| = (f \mid f)^{1/2}$.

Let $\mathcal{H}^2(D)$ be the subspace of $L^2(D)$ consisting of all square summable functions on D which are holomorphic on D. The space $\mathcal{H}^2(D)$, being a closed subspace of $L^2(D)$ is a separable Hilbert space; furthermore it has a reproducing kernel (the Bergman kernel function) $K: D \times D \to \mathbb{C}$. [cf. e.g. Helgason, 1].

Let $(e_\nu)_{\nu \in \mathbb{N}}$ be a complete orthonormal system in $\mathcal{H}^2(D)$. Then, for all $x,y \in D$,

$$K(x,y) = \sum_{\nu=0}^{+\infty} e_\nu(x)\overline{e_\nu(y)},$$

the convergence being uniform on compact subsets of $D \times D$.

Thus $K(y,x) = \overline{K(x,y)}$ and the function $K_y: x \mapsto K(x,y)$ is an element of $\mathcal{H}^2(D)$. Recall that the reproducing property of

K implies that $\|K_y\|^2 = K(y,y)$. Since D is bounded, for every $o \in D$, there is some $f \in \mathcal{H}^2(D)$ such that $f(o) \neq 0$. Thus $K(o,o) > 0$.

The evaluation at o is a continuous linear form $f \mapsto f(o)$ of norm one on $\mathcal{H}^2(D)$. Let $e_0 \in \mathcal{H}^2(D)$ be the vector representing the linear form:

$$(f|e_0) = f(o) \qquad (f \in \mathcal{H}^2(D)).$$

Then $\|e_0\| = 1$ and $e_0(o) \neq 0$. If $(e_1, e_2, \dots)$ is a complete orthogonal system in $e_0^{\perp}$, then $(e_0, e_1, \dots)$ is a complete orthonormal system in $\mathcal{H}^2(D)$. The domain D being bounded, for every $v \in \mathbb{C}^n \setminus \{0\}$, there is some $f \in \mathcal{H}^2(D)$ such that $f(o)=0$, $df(o)v \neq 0$. Hence we may assume that $de_1(o), \dots, de_n(o)$ are n linearly independent non-vanishing linear forms.

It is well known that the hermitian differential form

$$\sum_{\alpha,\beta=1}^{n} \frac{\partial^2 \log K}{\partial z^\alpha \partial \overline{z^\beta}} dz^\alpha \overline{dz^\beta}$$

defines on D a riemannian metric (the Bergman metric) which is invariant under the group Aut(D). For $z \in D$ we denote by $\langle\ ,\ \rangle_z$ the inner product defined by the Bergman metric at z:

$$\langle v_1, v_2 \rangle_z = \sum_{\alpha,\beta=1}^{n} \frac{\partial^2 \log K(z,z)}{\partial z^\alpha \partial \overline{z^\beta}} v_1^\alpha \overline{v_2^\beta}$$

for $v_1 = (v_1^1, \dots, v_1^n)$, $v_2 = (v_2^1, \dots, v_2^n)$ in $\mathbb{C}^n$.

We have, by a straightforward computation,

$$\sum_{\alpha,\beta=1}^{n} \frac{\partial^2 \log K(z,z)}{\partial z^\alpha \partial \overline{z^\beta}} dz^\alpha \overline{dz^\beta} = \frac{1}{2K(z,z)^2} \sum_{p,q=0}^{+\infty} |e_p(z)de_q(z) - e_q(z)de_p(z)|^2$$

for z in D. In particular

$$\sum_{\alpha,\beta=1}^{n} \left(\frac{\partial^2 \log K}{\partial z^\alpha \partial \overline{z^\beta}}\right)_{z=o} dz^\alpha \overline{dz^\beta} = \frac{1}{2K(o,o)} \sum_{p=1}^{+\infty} |de_p(o)|^2,$$

i.e.

(V.6.1) $$<v_1, v_2>_o = \frac{1}{K(o,o)} \sum_{p=1}^{+\infty} (de_p(o)v_1)\overline{(de_p(o)v_2)}.$$

Theorem V.6.1. [K.T. Hahn, 1]. *For any* $o \in D$ *and any* $v \in \mathbb{C}^n$

$$\gamma_D(o;v)^2 \leqslant \langle v,v\rangle_o .$$

Proof. To prove the theorem we will show that

(V.6.2) $$|df(o)\cdot v|^2 \leqslant \langle v,v\rangle_o$$

for all $f \in Hol(D,\Delta)$ for which $f(o) = 0$.

Let $a \in \mathcal{H}^2(D)$ be defined by $a(x) = f(x)K_0(x)$. Then

(V.6.3) $$\|a\|^2 = (a|a) = \int_D f(x)K_0(x)\overline{f(x)K_0(x)}\,d\omega(x) =$$

$$= \int_D |f(x)|^2|K_0(x)|^2 d\omega(x) \leqslant \|K_0\|^2 = K(o,o).$$

For $z \in D$, consider the function $x \mapsto \frac{K_z(x)}{K(z,z)}$ on D. Let $b(x) = \left(\overline{\partial}_z \frac{K_z(x)}{K(z,z)}\right)_{z=o}(v)$. If $v^1, \dots, v^n$ are the components of v, then

$$b(x) = \sum_{\beta=1}^{n} \overline{v^\beta} \left(\frac{\partial}{\partial \bar{z}^\beta} \frac{K_z(x)}{K(z,z)}\right)_{z=o}$$

Choosing a complete orthonormal system as above, we have by a direct computation,

$$\left(\frac{\partial}{\partial \bar{z}^\beta} K(x,z)\right)_{z=o} = \sum_{\nu=0}^{+\infty} e_\nu(x) \overline{\left(\frac{\partial e_\nu}{\partial z^\beta}\right)_o} ,$$

$$\left(\frac{\partial}{\partial \bar{z}^\beta} K(z,z)\right)_{z=o} = \sum_{\nu=0}^{+\infty} e_\nu(o) \overline{\left(\frac{\partial e_\nu}{\partial z^\beta}\right)_o} = e_0(o) \overline{\left(\frac{\partial e_0}{\partial z^\beta}\right)_o} ,$$

and therefore

$$\left(\frac{\partial}{\partial \bar{z}^\beta} \frac{K(x,z)}{K(z,z)}\right)_{z=o} = \frac{1}{K(o,o)^2}\left(K(o,o)\left(\frac{\partial K(x,z)}{\partial \bar{z}^\beta}\right)_{z=o} - K(x,o)\left(\frac{\partial K(z,z)}{\partial \bar{z}^\beta}\right)_{z=o}\right) =$$

$$= \frac{1}{K(o,o)^2}\left(|e_0(o)|^2 \sum_{\nu=0}^{+\infty} e_\nu(x) \overline{\left(\frac{\partial e_\nu}{\partial z_\beta}\right)_o} - |e_0(o)|^2 e_0(x) \overline{\left(\frac{\partial e_0}{\partial z^\beta}\right)_o}\right) =$$

$$= \frac{1}{K(o,o)} \sum_{\nu=1}^{+\infty} e_\nu(x) \overline{\left(\frac{\partial e_\nu}{\partial z^\beta}\right)_o} .$$

Thus

$$b(x) = \frac{1}{K(o,o)} \sum_{\nu=1}^{+\infty} e_\nu(x) \overline{(de_\nu(o)v)}$$

and, by (V.6.1),

$$\text{(V.6.4)} \qquad \|b\|^2 = \frac{1}{K(o,o)^2} \sum_{\nu=1}^{+\infty} |de_\nu(o)v|^2 = \frac{1}{K(o,o)} \langle v,v\rangle_o .$$

By the reproducing property of K

$$(a|b) = \sum_{\alpha=1}^{n} v^\alpha \left(fK_o \middle| \left(\frac{\partial}{\partial \bar{z}^\alpha} \frac{K_z}{K(z,z)}\right)_{z=o}\right) =$$

$$= \sum_{\alpha=1}^{n} v^\alpha \left(\frac{\partial}{\partial z^\alpha} \left(fK_z \middle| \frac{K_z}{K(z,z)}\right)\right)_{z=o} =$$

$$= \sum_{\alpha=1}^{n} v^\alpha \left(\frac{1}{K(z,z)} \frac{\partial}{\partial z^\alpha} (fK_z|K_z)\right)_{z=o} =$$

$$= \sum_{\alpha=1}^{n} v^\alpha \left(\frac{1}{K(z,z)} \frac{\partial}{\partial z^\alpha} f(z)K_z(z)\right)_{z=o} = \sum_{\alpha=1}^{n} \left(\frac{\partial f}{\partial z^\alpha}\right)_o v^\alpha = df(o)v.$$

The Schwarz inequality, together with (V.6.3) and (V.6.4) yield the conclusion.

Q.E.D.

Notes.

The notion of Carathéodory metric appears for the first time in [Carathéodory, 1 and 2] for domains in $\mathbb{C}^2$ (cf. also [Behnke-Thullen, 1, p.165]). The Carathéodory distance has been defined and thoroughly investigated for complex spaces in [Reiffen, 1 and 2]. The definition has been extended to domains in complex Banach spaces and applied to the proof of a fixed point theorem by C.J. Earle and R.S. Hamilton in [Earle-Hamilton, 1].

§ 2 develops the study of local properties of the Carathéodory metric, including - among other things - the fact that the Carathéodory metric is logarithmically plurisubharmonic (Theorem V.2.3). This fact was noticed by S. Kobayashi in [Kobayashi, 5, p. 400].

The fixed point theorem of C.J. Earle and R.S. Hamilton is proved in § 5 using the Kobayashi metric (cf. [Hervé, 4] for further generalizations). The main properties of this metric have been investigated by H.L. Royden in [Royden, 1] for domains in $\mathbb{C}^n$ and in [Royden, 2] for finite dimensional connected complex manifolds. In these two cases, Royden proved that the Kobayashi distance is the integrated form of the Kobayashi metric. Royden's proof goes over, with no substantial change, to domains in a Banach space (Theorem V.4.1) and even to domains in locally convex spaces (cf. [Vesentini, 4]). All this implies that in a hyperbolic domain the Kobayashi distance is an inner distance (for a direct proof of this fact, cf. [Kobayashi, 4]). A thorough report on inner distances is in [Rinow, 1].

On the other hand, the Carathéodory distance is in general

not inner, as it is shown at the end of § 4.

On hyperbolic domains the Kobayashi metric can not degenerate (Proposition V.1.9). This fact entails Theorem V.1.10, which says, roughly, that a family of holomorphic mappings of a hyperbolic domain D into itself, containing a holomorphic automorphism of D, cannot depend holomorphically on a complex parameter. This result and its proof extend, with no substantial change, to the case in which D is a connected hyperbolic manifold. A similar theorem was established for a bounded domain $D \subset \mathbb{C}^n$, by a quite different argument, in [Andreotti-Vesentini, 1, pp. 270-271].

According to Theorem V.6.1, the Bergman metric of a bounded domain in $\mathbb{C}^n$ dominates the Carathéodory metric [Hahn, 1].

CHAPTER VI

THE UNIT BALL IN A COMPLEX HILBERT SPACE

In this chapter H will be a complex Hilbert space with scalar product $(\mid)$ and norm $\| \|$, and B will denote the open unit ball of H.

Our purpose is to compute explicitly invariant distances and invariant metrics on B and to describe the group $\mathrm{Aut}(B)$ and the semigroup of all holomorphic isometries of the Carathéodory (and Kobayashi) distance.

All the results in the present chapter find their motivations in the classical, familiar case $H = \mathbb{C}$. Thus we shall assume throughout the chapter $\dim_{\mathbb{C}} H > 1$. For the case $H = \mathbb{C}$, see Appendix A.

We begin by considering the group $\mathrm{Aut}(B)$.

§ 1. Automorphisms of the unit ball.

For any $b \in B$, let $\alpha(b) = \sqrt{1-\|b\|^2}$ and let T_b be the linear map $H \to H$ expressed by

(VI.1.1) $$T_b(x) = \frac{(x \mid b)}{1 + \alpha(b)} b + \alpha(b) x \, .$$

Note that

(VI.1.2) $$T_b(b) = b$$

and $T_0(x) = x$ for all $x \in H$. Let $b \in B\setminus\{0\}$ and $u = \frac{1}{\|b\|}b$. For any $x \in H$, we write $x = \zeta u + z$, where $\zeta = (x|u)$ and $z = x - (x|u)u$. Then

(VI.1.3) $$T_b(x) = \frac{\zeta\|b\|^2}{1+\alpha(b)}u + \alpha(b)\zeta u + \alpha(b)z = \zeta u + \alpha(b)z,$$

and therefore

(VI.1.4) $$\|T_b(x)\|^2 = |\zeta|^2 + \alpha(b)^2\|z\|^2 = |(x|u)| + (1-\|b\|^2)\|x-(x|u)u\|^2.$$

Hence $\|T_b(x)\| \leqslant \|x\|$, and, by (VI.1.2), $\|T_b\| = 1$. By (VI.1.2)

(VI.1.5) $$T_b \circ T_b(x) = \frac{(x|b)}{1+\alpha(b)}b + \alpha(b)T\ (x) = (x|b)b + \alpha(b)^2 x.$$

For any fixed $b \in B\setminus\{0\}$ the function $x \mapsto \frac{1}{1-(x|b)}$ is holomorphic on the open set $H\setminus\{x \in H: (x|b) < 1\}$. In particular, it is holomorphic on the open ball $B(0, \frac{1}{\|b\|}) = \{x \in H: \|x\| < \frac{1}{\|b\|}\}$. For any $b \in B$, let $f_b: B(0, \frac{1}{\|b\|}) \to H$ be defined by

$$f_b(x) = T_b\left(\frac{1}{1-(x|b)}(x-b)\right).$$

Then $f_b \in Hol(B_{\frac{1}{\|b\|}}, H)$. For $b=0$, $f_0(x) = x$ for all $x \in H$.

Let $b \in B\setminus\{0\}$, and let $u = \frac{1}{\|b\|}b$. Writing, as before, $x = \zeta u + z$ with $\zeta = (x|u)$, (VI.1.3) yields, for all $x \in B_{\frac{1}{\|b\|}}$,

(VI.1.6) $$f_b(x) = \frac{1}{1-(x|b)}((x|u) - \|b\|)u + \alpha(b)z).$$

Since

$$1 - \left|\frac{\zeta - \|b\|}{1-\|b\|\zeta}\right|^2 = \frac{(1-\|b\|^2)(1-|\zeta|^2)}{|1-\|b\|\zeta|^2}$$

and since, by (VI.1.6),

(VI.1.7) $$\|f_b(x)\|^2 = \frac{1}{|1-\|b\|\zeta|^2}(|\zeta-\|b\||^2+\alpha(b)^2\|z\|^2),$$

then

(VI.1.7') $$1-\|f_b(x)\|^2 = \frac{(1-\|b\|^2)(1-|\zeta|^2-\|z\|^2)}{|1-\|b\|\zeta|^2} = \frac{(1-\|b\|^2)(1-\|x\|^2)}{|1-(x|b)|^2},$$

whence $f_b(B) \subset B$.

We prove now that, for every $x \in B$,

(VI.1.8) $$f_{-b}\circ f_b(x) = x.$$

Let $x' = f_b(x)$. First of all, by (VI.1.6)

$$(x'|b) = \frac{1}{1-(x|b)}((x|b)-\|b\|)(u|b) = \frac{(x|b)-\|b\|^2}{1-(x|b)},$$

and therefore

$$1+(x'|b) = \frac{1-\|b\|^2}{1-(x|b)}.$$

Furthermore, by (VI.1.2) and (VI.1.5),

$$T_b(x'+b) = \frac{1}{1-(x|b)}(T_b\circ T_b(x)-b)+b =$$

$$= \frac{1}{1-(x|b)}((x|b)b+\alpha(b)^2x-b)+b = \frac{1-\|b\|^2}{1-(x|b)}x,$$

and therefore

$$x = \frac{1-(x|b)}{1-\|b\|^2}T_b(x'+b) = \frac{1}{1+(x'|b)}T_b(x'+b).$$

Being $T_b = T_{-b}$, we have then

$$x = f_{-b}(x'),$$

proving thereby (VI.1.8) and also

$$f_b\circ f_{-b}(x) = x \qquad (x \in B).$$

That proves

Lemma VI.1.1. *For every* $b \in B$, *the restriction of* f_b *to* B *is a holomorphic automorphism of* B.

For the sake of simplicity we denote $f_{b|B}$ by f_b.

Hence $f_b \in \mathrm{Aut}(B)$, and

$$f_b^{-1} = f_{-b} \,.$$

<u>Lemma VI.1.2</u>. *A linear map* $U \in L(H)$ *belongs to* Aut B *if, and only if,* U *is unitary.*

<u>Proof</u>. The map U must be invertible, and furthermore

$$\|Ux\| = \|x\| \qquad \text{for all } x \in H.$$

In view of the identity

$$(x|y) = \frac{1}{4}\{\|x+y\|^2 - \|x-y\|^2 + i(\|x+iy\|^2 - \|x-iy\|^2)\},$$

that is equivalent to

$$(Ux|Uy) = (x|y) \qquad \text{for all } x,y \in H.$$

Q.E.D.

<u>Theorem VI.1.3</u>. *For any* $F \in \mathrm{Aut}(B)$, *there exist a vector* $b \in B$ *and a unitary transformation* U *of* H *such that*

$$F = U \circ f_b \,.$$

<u>Proof</u>. Let $b = F^{-1}(0)$. Being $f_b(b) = 0$, then $f_b^{-1}(0) = b$, and the automorphism $F \circ f_b^{-1}$ leaves 0 invariant. By Proposition III.2.2, $F \circ f_b^{-1}$ is linear, and therefore, by Lemma VI.1.2, unitary.

Q.E.D.

Since f_b is holomorphic on $B(0, \frac{1}{\|b\|})$, we have also

<u>Proposition VI.1.4</u>. *Any* $g \in \mathrm{Aut}(B)$ *is the restriction to* B *of a holomorphic map of* $B(0, \frac{1}{\|g^{-1}(0)\|})$ *into* H.

Being $f_b(b) = 0$, the family $\{f_b\colon b \in B\}$ acts transitively in B. Hence

<u>Proposition VI.1.5</u>. *The unit ball* B *of a complex Hilbert*

space H *is homogeneous.*

Lemma VI.1.6. *For every* $g \in \mathrm{Aut}(B)$ *there exists a continuous linear isomorphism* h *of* $\mathbb{C} \times H$ *onto* $\mathbb{C} \times H$ *such that, for any* $x \in B$, *the image of the complex line through* $(1,x)$ *is the complex line through* $(1,g(x))$, i.e., *for any* $x \in B$ *there is* $\sigma \in \mathbb{C}$ *for which*

$$h(1,x) = \sigma(1,g(x)).$$

Proof. By Theorem VI.1.3, $g = U \circ f_b$, for a suitable $b \in B$ and a suitable unitary operator U. The map

$$h: (\zeta,x) \mapsto (\zeta-(x|b),\ U \circ T_b(x-\zeta b)) \qquad (\zeta \in \mathbb{C},\quad x \in B)$$

is the restriction to $\mathbb{C} \times B$ of a linear continuous map h of $\mathbb{C} \times H$ into $\mathbb{C} \times H$. Setting $(\zeta',x') = h(\zeta,x)$, then

$$\zeta' = \zeta - (x|b), \qquad x' = U \circ T_b(x-\zeta b).$$

Since, by (VI.1.1), (VI.1.2),

$$(U^{-1}x'|b) = \frac{(x|b)}{1+\alpha(b)}\|b\|^2 + \alpha(b)(x|b) - \zeta\|b\|^2 = (x|b) - \zeta\|b\|^2,$$

then

$$\zeta = \zeta' + (x|b) = \zeta' + (U^{-1}x'|b) + \zeta\|b\|^2, \tag{VI.1.9}$$

i.e.

$$\zeta = \frac{1}{\alpha(b)^2}(\zeta' + (U^{-1}x'|b)). \tag{VI.1.10}$$

Furthermore, by (VI.1.2), (VI.1.5), (VI.1.9),

$$T_b(U^{-1}x' + \zeta' b) = T_b(T_b(x-\zeta b)+\zeta' b) = T_b \circ T_b(x) + (\zeta'-\zeta)b =$$

$$= \alpha(b)^2 x,$$

and therefore

$$x = \frac{1}{\alpha(b)^2} T_b(U^{-1}x' + \zeta' b). \tag{VI.1.11}$$

Formulas (VI.1.10) and (VI.1.11) complete the proof of the lemma.

Q.E.D.

Recall that an *affine subspace* L of H is a set of the form $L = x_0 + F$ where $x_0 \in H$ and F is a vector subspace of H. Clearly L is closed if, and only if, F is closed.

The set $(\{1\} \times L) \subset \mathbb{C} \times H$ is the intersection of $\{1\} \times H$ with a vector subspace of $\mathbb{C} \times H$. Viceversa, the intersection of $\{1\} \times H$ with a vector subspace of $\mathbb{C} \times H$ defines an affine subspace of H. The above lemma implies that any $g \in \operatorname{Aut}(B)$ preserves affine subspaces. More specifically we have

Theorem VI.1.7. *Let* $g \in \operatorname{Aut}(B)$. *If* $L \cap B \neq \emptyset$, *then* $g(L \cap B)$ *is the intersection of* B *with an affine subspace of* H.

We shall now consider briefly holomorphic maps of B into B.

Proposition VI.1.8. *Let* $f \in \operatorname{Hol}(B,B)$ *with* $f(0) = 0$. *The set*

$$M = \{x \in B: \|f(x)\| = \|x\|\}$$

is the intersection of B *with a closed subspace* L *of the Hilbert space* H. *The restriction of* f *to* M *is the restriction to* M *of a partial linear isometry of* H, *whose initial space is* L, *and whose final space is* $df(0)L$.

Proof. Since every point with norm one in H is a real extreme point of $\overline{B}$ (Remark 1, § III.1), then, by Theorem III.2.3, for every $x \in M$ and all $\zeta \in \mathbb{C}$ with $|\zeta| < \frac{1}{\|x\|}$, we have

$$f(\zeta x) = \zeta f(x) \qquad \text{for all } |\zeta| < \frac{1}{\|x\|} . \tag{VI.1.12}$$

Let $\sum_{q=1}^{+\infty} P_q$ be the power series expansion of f in M.

Then, by (VI.1.12):

(VI.1.13) $P_q(x) = 0$ for $q=2,3,\ldots$ and for all $x \in M$, and so
$f(\zeta x) = \zeta P_1(x)$

and

(VI.1.14) $\|P_1(x)\| = \|x\|$ on M.

Thus if $x \in M$, then (VI.1.14) holds. We prove now that the converse holds. We write $f = P_1+g$, where $g \in \mathrm{Hol}(B,H)$ is defined by

$$g(y) = \sum_{q=2}^{+\infty} P_q(y) \qquad (y \in B).$$

First of all, by the Cauchy inequalities (Proposition II.3.6),

(VI.1.15) $\|P_1\| \leqslant 1$.

By the Schwarz lemma (Theorem III.2.3)

(VI.1.16) $\|f(\zeta x)\|^2 = \|P_1(\zeta x)\|^2 + \|g(\zeta x)\|^2 + 2\mathrm{Re}(g(\zeta x)|P_1(\zeta x)) \leqslant$
$\leqslant |\zeta|^2\|x\|^2$

for all $x \in B$ and all $|\zeta| < \frac{1}{\|x\|}$. Thus, if x satisfies (VI.1.14), then

$$\|g(\zeta x)\|^2 + 2\mathrm{Re}(g(\zeta x)|P_1(\zeta x)) \leqslant 0$$

for all $|\zeta| < \frac{1}{\|x\|}$, and therefore $\mathrm{Re}(g(\zeta x)|P_1(\zeta x)) \leqslant 0$ for all $|\zeta| < \frac{1}{\|x\|}$. Being

$$(g(\zeta x)|P_1(\zeta x)) = |\zeta|^2\{\zeta(P_2(x)|P_1(x)) + \zeta^2(P_3(x)|P_1(x)) + \ldots\},$$

we conclude that $(g(\zeta x)|P_1(\zeta x)) = 0$ for all ζ, and therefore $(P_q(x)|P_1(x)) = 0$ for $q=2,3,\ldots$; (VI.1.16) yields

$$\|f(x)\|^2 = \|P_1(x)\|^2 + \|g(x)\|^2 \leqslant \|x\|^2 ,$$

whence, by (VI.1.14), $g(x) = 0$; so $\|f(x)\| = \|P_1(x)\|$, and $\|f(x)\| = \|x\|$ again by (VI.1.14). In conclusion we have

$$\text{(VI.1.17)} \qquad M = \{x \in B : \|P_1(x)\| = \|x\|\}.$$

Now, let $x_1, x_2 \in M$. For $\zeta_1, \zeta_2 \in \mathbb{C}$ we have, by (VI.1.14),

$$\|\zeta_1 x_1 + \zeta_2 x_2\|^2 = |\zeta_1|^2\|x_1\|^2 + |\zeta_2|^2\|x_2\|^2 + 2\text{Re}(\zeta_1 x_1 | \zeta_2 x_2),$$

$$\|P_1(\zeta_1 x_1 + \zeta_2 x_2)\|^2 = |\zeta_1|^2\|P_1(x_1)\|^2 + |\zeta_2|^2\|P_1(x_2)\|^2 +$$
$$+ 2\text{Re}(P_1(\zeta_1 x_1) | P_1(\zeta_2 x_2)) = |\zeta_1|^2\|x_1\|^2 + |\zeta_2|^2\|x_2\|^2 +$$
$$+ 2\text{Re}(P_1(\zeta_1 x_1) | P_1(\zeta_2 x_2)).$$

Since, by (VI.1.15),

$$\|P_1(\zeta_1 x_1 + \zeta_2 x_2)\| \leqslant \|\zeta_1 x_1 + \zeta_2 x_2\| ,$$

then

$$\text{Re}(P_1(\zeta_1 x_1) | P_1(\zeta_2 x_2)) \leqslant \text{Re}(\zeta_1 x_1 | \zeta_2 x_2),$$

for all $\zeta_1, \zeta_2 \in \mathbb{C}$. Hence $(P_1(x_1) | P_1(x_2)) = (x_1 | x_2)$, and therefore

$$\|P_1(\zeta_1 x_1 + \zeta_2 x_2)\| = \|\zeta_1 x + \zeta_2 x_2\|$$

for all $\zeta_1, \zeta_2 \in \mathbb{C}$. By (VI.1.17) $\zeta_1 x_1 + \zeta_2 x_2 \in M$ whenever ζ_1, ζ_2 are such that $\zeta_1 x_1 + \zeta_2 x_2 \in B$. Hence M is the intersection of B with a linear subvariety L of H. Since P_1 is continuous, condition (VI.1.14) implies that M is closed.

The restriction of f to M is P_1, which by (VI.1.14) is a linear isometry of L.

Let P be the orthogonal projection of H whose image is L. The linear map $P_1 \circ P$ is a partial isometry of H whose initial space is L and final space is $P_1(L)$. Since $\|P_1 \circ P\| = 1$, then $P_1 \circ P \in \text{Hol}(B,B)$. Finally $P_1 \circ P(x) = P_1(x) = f(x)$ for all $x \in M$.

Q.E.D.

<u>Problem</u>. Does Proposition VI.1.8 hold for more general domains in complex Banach spaces?

§ 2. <u>Invariant distances and invariant metrics</u>.

Since B is homogeneous, Lemma V.1.5 yields

<u>Theorem VI.2.1</u>. *The Carathéodory and Kobayashi differential metrics on* B *coincide.*

For any $b \in B$ and $v \in H$, we have, by Lemma V.1.5,

$$\gamma_B(b;v) = \gamma_B(f_b(b);df_b(b)v) = \gamma_B(0;df_b(b)v) = \|df_b(b)v\|. \tag{VI.2.1}$$

By (VI.1.1), for all $x \in B$,

$$f_b(x) = \frac{1}{1-(x|b)}(T_b(x)-b) = \frac{1}{1-(x|b)}\left(\frac{(x|b)}{1+\alpha(b)}b + \alpha(b)x-b\right).$$

Setting $x = b+y$,

$$f_b(x) = f_b(b+y) = \frac{1}{1-\|b\|^2-(y|b)}\left(\frac{\|b\|^2+(y|b)}{1+\alpha(b)}b + \alpha(b)b + \alpha(b)y - b\right) =$$

$$= \frac{1}{\alpha(b)^2}\,\frac{1}{1-\dfrac{(y|b)}{\alpha(b)^2}}\,T_b(y)\,.$$

For $\left|\dfrac{(y|b)}{\alpha(b)^2}\right| < 1$ - hence, *a fortiori*, for $\|y\| < \dfrac{\alpha(b)^2}{\|b\|}$ -

$$f_b(b+y) = \frac{1}{\alpha(b)^2}\left(1 + \frac{(y|b)}{\alpha(b)^2} + \left(\frac{(y|b)}{\alpha(b)^2}\right)^2 + \dots\right)T_b(y)$$

so that

$$df_b(b) = \frac{1}{1-\|b\|^2}\,T_b\,.$$

For $b \in B\setminus\{0\}$, setting $u = \frac{1}{\|b\|}b$, and $v = (v|u)u+z$, then, by (VI.1.3),

$$T_b(v) = (v|u)u + \alpha(b)z,$$

so that

$$\|T_b(v)\|^2 = |(v|u)|^2 + \alpha(b)^2\|z\|^2 = |(v|b)|^2 + (1-\|b\|^2)\|v\|^2 .$$

Hence

$$\|df_b(b)v\|^2 = \frac{1}{(1-\|b\|^2)^2}\{|(v|b)|^2 + (1-\|b\|^2)\|v\|^2\},$$

and (VI.2.1) becomes

$$\gamma_B(b;v) = \frac{1}{1-\|b\|^2}(|(v|b)|^2 + (1-\|b\|^2)\|v\|^2)^{1/2} .$$

For every $b \in B$, consider the hermitian sesquilinear form

$$\text{(VI.2.2)} \qquad \{v_1,v_2\}_b = \frac{1}{(1-\|b\|^2)^2}((v_1|b)\overline{(v_2|b)} + (1-\|b\|^2)(v_1|v_2)) .$$

Being

$$\{v,v\}_b = \gamma_B(b;v)^2,$$

$\{\ ,\ \}_b$ is positive definite, and continuous. By the Riesz representation theorem, there is a unique hermitian continuous operator C_b on H such that

$$\{v_1,v_2\}_b = (C_b v_1|v_2) .$$

The operator C_b can be easily guessed from (VI.2.2). It is given by

$$C_b(v) = \frac{1}{(1-\|b\|^2)^2}((v|b)b + (1-\|b\|^2)v) \qquad (v \in H) .$$

For $b = 0$, C_b is the identity: $C_0 = I$. Let $b \in B\setminus\{0\}$, and let P_b be the orthogonal projection onto the complex line defined by b,

$$P_b : v \mapsto \frac{(v|b)}{\|b\|^2} b .$$

Then

$$C_b = \left(\frac{\|b\|}{1-\|b\|^2}\right)^2 P_b + \frac{1}{1-\|b\|^2} I .$$

The spectrum of C_b consists of the two eigenvalues $\frac{1}{1-\|b\|^2}$ and $\frac{1}{(1-\|b\|^2)^2}$, whose eigenspaces are, respectively,

$b^{\perp}$ and the complex line defined by b . The numerical range of C_b is therefore the closed interval $[\frac{1}{1-\|b\|^2}, \frac{1}{(1-\|b\|^2)^2}]$.

For $b \in B$ and $v \in H$, consider the function

$$\zeta \mapsto \log \frac{1}{1-\|b+\zeta v\|^2},$$

which is of class C^{∞} in a neighborhood of 0 in $\mathbb{C}$. We have

$$\frac{\partial}{\partial\bar{\zeta}} \log \frac{1}{1-\|b+\zeta v\|^2} = - \frac{1}{1-\|b+\zeta v\|^2},$$

and therefore

$$\left(\frac{\partial^2}{\partial\zeta\partial\bar{\zeta}} \log \frac{1}{1-\|b+\zeta v\|^2}\right)_{\zeta=0} = \frac{|(v|b)|^2}{(1-\|b\|^2)^2} + \frac{\|v\|^2}{1-\|b\|^2},$$

i.e.

$$\left(\frac{\partial^2}{\partial\zeta\partial\bar{\zeta}} \log \frac{1}{1-\|b+\zeta v\|^2}\right)_{\zeta=0} = \gamma_B(b;v)^2 .$$

If H has finite dimension, the left-hand side defines the Bergman differential metric of B . Hence this metric coincides with γ_B and κ_B . Kobayashi proved in [Kobayashi, 3, p. 52] that Carathéodory's and Kobayashi's *distances* on a bounded symmetric domain in $\mathbb{C}^n$ coincide; his argument yields the same conclusion for the corresponding differential metrics.

We consider now the invariant distances c_B and k_B.

Since B is homogeneous, Theorem IV.1.8 and IV.2.6 yield

Theorem VI.2.2 *The Carathéodory distance* c_B *coincides with the Kobayashi distance* k_B *, and is complete.*

For $x,y \in B$, the function $f_y \in \mathrm{Aut}(B)$ is such that $f_y(y) = 0$. Hence, by Theorem IV.1.8,

$$c_B(x,y) = c_B(f_y(x),0) = \omega(0,\|f_y(x)\|) .$$

If $y \in B\setminus\{0\}$, then, setting $u = \frac{1}{\|y\|}y$, $z = x - (x|u)u$, we have $z \perp u$, and, by (VI.1.7'),

$$\|f_y(x)\|^2 = \frac{1}{|1-(x|y)|^2}\{|1-(x|y)|^2 - (1-\|x\|^2)(1-\|y\|^2)\}.$$

Thus

$$c_B(x,y) = \omega(0, \frac{1}{|1-(x|y)|}(|1-(x|y)|^2 - (1-\|x\|^2)(1-\|y\|^2))^{1/2}).$$

Let L be a closed subspace of H and let P be the orthogonal projection having range L. P defines a holomorphic map of B onto $B \cap L$, which is the identity on $B \cap L$. On the other hand, the identity map $L \to H$ defines a holomorphic map of $B \cap L$ into B. Applying Proposition IV.1.2 to these two maps, we obtain, for $x_1, x_2 \in B \cap L$,

$$k_{B\cap L}(x_1,x_2) \geqslant k_B(x_1,x_2) \geqslant k_{B\cap L}(Px_1,Px_2) = k_{B\cap L}(x_1,x_2),$$

i.e.

$$k_{B\cap L}(x_1,x_2) = k_B(x_1,x_2),$$

and similarly

$$c_{B\cap L}(x_1,x_2) = c_B(x_1,x_2),$$

Now, let F be a closed affine subspace of H such that $F \cap B \neq \emptyset$. Choosing any point $x_0 \in F \cap B$, and any $g \in \mathrm{Aut}(B)$ such that $g(x_0) = 0$, $g(F \cap B)$ is the intersection of B with a vector subspace L of H. Furthermore, $g|_{F\cap B}$ is a bi-holomorphic map of $F \cap B$ onto $L \cap B$. Thus we have

$$k_{B\cap F}(x_1,x_2) = k_B(x_1,x_2) = c_B(x_1,x_2) = c_{B\cap F}(x_1,x_2)$$

for all $x_1, x_2 \in F \cap B$. We have proved

Lemma VI.2.3. *The Carathéodory and Kobayashi distances* $c_{B\cap F}$ *and* $k_{B\cap F}$ *coincide with* $c_B = k_B$ *on* $B \cap F$.

Similarly one can prove that the Carathéodory and Kobayashi

metrics $\gamma_{B\cap F}$ and $\kappa_{B\cap F}$ coincide with the Carathéodory and Kobayashi metrics γ_B and κ_B on $B \cap F$.

As a consequence of Proposition VI.1.7 we prove now

Proposition VI.2.4. *Let* $f \in \mathrm{Hol}(B,B)$. *For any* $x_0 \in B$, *set*

$$S_f(x_0) = \{x \in B: c_B(f(x),f(x_0)) = c_B(x,x_0)\}.$$

There exist: a partial isometry Q *of* H, *and two holomorphic automorphisms* g_1, g_2 *of* B *such that, denoting by* L *and* N *the initial and final spaces of* Q, *we have*

$$g_1(0) = x_0,$$

$$g_1(L \cap B) = S_f(x_0),$$

$$g_2(N \cap B) = f(S_f(x_0)),$$

$$f|_{S_f(x_0)} = g_2 \circ Q \circ g_1^{-1}|_M,$$

where $M = \{x \in B: \|h(x)\| = \|x\|\}$.

Proof. Let $g_1, g_2 \in \mathrm{Aut}(B)$ be such that $x_0 = g_1(0)$, $f(x_0) = g_2(0)$. Let $h = g_2^{-1} \circ f \circ g_1 \in \mathrm{Hol}(B,B)$. Then $h(0) = 0$. By Proposition VI.1.3, the set M is the intersection of B with a closed subspace L of H; furthermore there exists a partial isometry Q of H, whose initial space is L and the final space is $dh(0)(L)$, such that $Q|_M = h|_M$.

Now

$$x \in S_f(x_0) \Leftrightarrow c_B(f(x),f(x_0)) = c_B(x,x_0) \Leftrightarrow c_B(g_2^{-1}(f(x)),0) =$$

$$= c_B(g_1^{-1}(x),0) \Leftrightarrow c_B(g_2^{-1} \circ f \circ g_1(g_1^{-1}(x)),0) = c_B(g_1^{-1}(x),0) \Leftrightarrow$$

$$\Leftrightarrow c_B(h(g_1^{-1}(x)),0) = c_B(g_1^{-1}(x),0) \Leftrightarrow \|h(g_1^{-1}(x))\| = \|g_1^{-1}(x)\| \Leftrightarrow$$

$$\Leftrightarrow g_1^{-1}(x) \in M \Leftrightarrow x \in g_1(M), \quad \text{i.e.} \quad S_f(x_0) = g_1(M).$$

Furthermore, for $x \in S_f(x_0)$, $f(x) = g_2 \circ h \circ g_1^{-1}(x) =$

$= g_2 \circ Q \circ g_1^{-1}(x)$.

Q.E.D.

Corollary VI.2.5. *$S_f(x_0)$ and $f(S_f(x_0))$ are intersections of B with closed affine subspaces of H.*

The following theorem characterizes the Kobayashi distance on the unit ball B of a complex Hilbert space H.

For $x \in B$, $r > 0$, we set $B(x,r) = \{y \in H: \|x-y\| < r\}$ and $B_d(x,r) = \{y \in H: d(x,y) < r\}$.

Theorem VI.2.6. *Let d be a distance on B such that:*

i) *d is topologically equivalent to the norm;*

ii) *$d(f(x),f(y)) \leq d(x,y)$ for all $f \in \mathrm{Hol}(B,B)$ and all x,y in B;*

iii) *$B_d(x,r+r') = B_d(B_d(x,r),r')$ for all $x \in B$, $r > 0$, $r' > 0$.*

Then, up to a positive constant factor, $d = k_B$.

Remarks. 1. By Theorem IV.2.2, Proposition IV.1.2, and Lemma IV.1.12, the Kobayashi distance satisfies i),ii) and iii).

2. By ii) the holomorphic automorphisms of B are surjective isometries for d.

Proof. We divide the proof into several steps.

A. Note first that for all $0<t\leq 1$, the holomorphic function $x \mapsto tx$ maps B into B. For all $s > 0$, the image of $B_d(0,s)$ is contained in $B_d(0,s)$.

If $x \in B_d(0,r)$, then $\overline{B(0,\|x\|)}^{\|\ \|} \subset B_d(0,r)$. In fact, for any y $\overline{B(0,\|x\|)}^{\|\ \|}$ there is some $0\leq t\leq 1$ and some unitary operator U such that $y = tUx$. The holomorphic function $z \mapsto tUz$ belongs to $\mathrm{Hol}(B,B)$ and maps x into y and $B_d(0,r)$

into itself.

B. Let $s_0 > 0$, and let

(VI.2.3) $\quad r_0 = \sup\{r: B(0,r) \subset B_d(0,s_0)\}.$

We denote by $S(x,r)$ and $S_d(x,r)$ the spheres with center x and radius r for the norm and for d, i.e.,

$$S(x,r) = \{y \in B: \|x-y\| = r\},$$

$$S_d(x,r) = \{y \in B: d(x,y) = r\},$$

and we prove that

(VI.2.4) $\quad S(0,r_0) = S_d(0,s_0).$

Note that $B(0,r_0) \subset B\ (0,s_0)$. If $r_0 = 1$, there is nothing to prove. Thus we assume $r_0 < 1$.

We show first that

$$x \in S(0,r_0) \Rightarrow d(0,x) = s_0.$$

If $d(0,x) < s_0$, i.e. $x \in B_d(0,s_0)$, then, by i), there is some $t > 0$ such that

$$B(x,t) \subset B_d(0,s_0).$$

Hence there is some $y \in B(x,t)$ with $\|y\| = \|x\| + \frac{t}{2} = r_0 + \frac{t}{2}$. Therefore

$$B(0,r_0 + \tfrac{t}{2}) \subset \overline{B(0,r_0 + \tfrac{t}{2})}^{\|\ \|} \subset B_d(0,r_0).$$

Contradiction.

If $d(0,x) > s_0$, there is $s > 0$ such that

$$B_d(x,s) = B \setminus \overline{B_d(0,s_0)}^{d} \subset B \setminus \overline{B(0,r_0)}^{\|\ \|},$$

and that is a contradiction. Thus $d(0,x) = s_0$, i.e.

$$S(0,r_0) \subset S_d(0,s_0).$$

Now, let $x \in S_d(0,s_0)$. Then $\|x\| \geqslant r_0$ (since, otherwise,

$x \in B(0,r_0) \subset B_d(0,s_0))$. Suppose that $\|x\| > r_0$. There is some $y \in B_d(0,s_0)$ with $\|y\| > r_0$. Since, by A,

$$\overline{B(0,\|y\|)}^{\|\ \|} \subset B_d(0,s_0),$$

that contradicts the definition. Thus $\|x\| = r_0$, and this completes the proof of (VI.2.4).

If $x \in B_d(0,s_0)$, then, by A,

$$\overline{B(0,\|x\|)}^{\|\ \|} \subset B_d(0,s_0)$$

i.e., $\|x\| \leqslant r_0$, and thus, by (VI.2.4), $\|x\| < r_0$.

Hence

$$B_d(0,s_0) = B(0,r_0).$$

In conclusion, there is a continuous strictly increasing function $\varphi: [0,1) \to \mathbb{R}_+$ such that

$$B(0,r) = B_d(0,\varphi(r)).$$

Let $\psi = \varphi^{-1}: \varphi([0,1)) \to [0,1)$.

C. Let $b \in B$ and $r > 0$. We will first prove that $B_d(b,r)$ is convex. Since $f_b(b) = 0$, then

$$x \in B_d(b,r) \Leftrightarrow f_b(x) \in B_d(0,r) \Leftrightarrow \|f_b(x)\| < \psi(r),$$

i.e.,

$$B_d(b,r) = B_d(f_{-b}(0),r) = f_{-b}(B_d(0,r)) = f_{-b}(B(0,\psi(r))). \tag{VI.2.5}$$

Let $x_1, x_2 \in B_d(b,r)$. The line segment $L = \{tx_1+(1-t)x_2 : 0 \leqslant t \leqslant 1\}$ belongs to the intersection of B with the complex affine line passing through x_1, x_2. The image by f_b of this intersection is the unit disc cut on a complex line by B. The image $f_b(L)$ of the line segment L belongs either to a diameter of this unit disc or to a circle orthogonal to the boundary of the unit disc. Hence

$$f_b(L) \subset B(0,\psi(r))$$

and that proves that $B_d(b,r)$ is convex.

We prove now that $B_d(b,r)$ is *symmetric about the line* $\{tb: t \in \mathbb{R}\}$. For this purpose we will show that, if U is any unitary operator having b as an eigenvector, with eigenvalue 1, then

$$U(B_d(b,r)) = B_d(b,r).$$

In view of (VI.2.5) we need only show that $x \in B_d(0,r)$ if, and only if,

(VI.2.6) $$f_b \circ U \circ f_{-b}(x) \in B_d(0,r).$$

Since $f_b \circ U \circ f_{-b} \in \text{Aut}(B)$, and

$$f_b \circ U \circ f_{-b}(0) = f_b \circ U_b(b) = f_b(b) = 0,$$

then, by Lemma VI.1.2, $f_b \circ U \circ f_{-b}$ is a unitary operator. Thus, by (VI.2.5), $f_b \circ U \circ f_{-b}(B_d(0,r)) = B_d(0,r)$, and therefore (VI.2.6) holds.

D. We prove now that, given $r>0$, $r'>0$, with $r+r' < 1$,

(VI.2.7) $$d(0,(r+r')u) = d(0,ru) + d(ru,(r+r')u).$$

Let $t = d(0,ru) = \varphi(r)$, $t+t' = d(0,(r+r')u) = \varphi(r+r')$. Then $t'>0$, and, by iii),

$$B_d(0,t+t') = B_d(B_d(0,t),t').$$

Hence there is some $y \in \overline{B_d(0,t)}$ such that

$$d(y,(r+r')u) \leqslant t'.$$

The triangle inequality yields

$$t+t' = d(0,(r+r')u) \leqslant d(0,y) + d(y,(r+r')u \leqslant t+t'.$$

Thus

(VI.2.8) $$d(0,(r+r')u) = d(0,y) + d(y,(r+r')u),$$

$$d(0,y) = t, \quad d(y,(r+r')u) = t',$$

and therefore, setting

$$K = \overline{B_d(0,t)} \cap \overline{B_d((r+r')u,t')},$$

we must have $y \in K$.

By C, K is convex and symmetric about the line $L = \{tu: t \in \mathbb{R}\}$. Thus, if $K = \{y\}$, then $y = ru$, in which case (VI.2.8) becomes (VI.2.7). In order to establish the latter formula, we shall show that indeed $K = \{y\}$. For any $z \in K$, the triangle inequality $t+t' = d(0,(r+r')u) \leqslant d(0,z) + d(z,(r+r')u \leqslant \leqslant t+t'$ yields $d(0,e) = t$. If $z \neq ru$, K contains a small disc perpendicular to the line L and passing through z; more specifically, there is a vector $v \in L$ and there is some $\rho > 0$ such that

$$\{v + \zeta(z-v): |\zeta| \leqslant \rho\} \subset K.$$

But all the points of this disc would have distance t from 0, contradicting the fact that all points with norm $\psi(t)$ are extreme points of the closed ball $B(0,\psi(t))$. Thus $K = \{y\}$.

E. Choose $r > 0$, $r' > 0$, $r+r' < 1$. Since $f_{ru}(ru) = 0$, then, by ii),

$$d(ru,(r+r')u) = d(0,f_{ru}(r+r')u).$$

Since, by (VI.1.6),

$$f_{ru}((r+r')u) = \frac{1}{1-r(r+r')}((r+r')-r)u = \frac{r'}{1-r(r+r')}u,$$

then

$$d(ru,(r+r')u) = d(0, \frac{r'}{1-r(r+r')} u) = \varphi\left(\frac{r'}{1-r(r+r')} \right),$$

and (VI.2.7) becomes

$$\text{(VI.2.9)} \qquad \varphi(r+r') - \varphi(r) = \varphi\left(\frac{r'}{1-r(r+r')}\right).$$

We prove now that φ has derivative at all points of $[0,1)$. We prove first that φ has derivative at $r=0$. Since φ is strictly increasing, then φ has derivative almost everywhere on $[0,1)$. Let $r \in [0,1)$ be such that $\varphi'(r)$ exists. By (VI.2.9),

$$\text{(VI.2.10)} \qquad \frac{\varphi\left(\frac{r'}{1-r(r+r')}\right)}{\frac{r'}{1-r(r+r')}} = \frac{\varphi(r+r') - \varphi(r')}{r'}\,(1-r(r+r')).$$

Passing to the limit as $r' \to 0$, we see that $\varphi'(0)$ exists. Now, for any $r \in [0,1)$, (VI.2.10) shows that the limit of $\frac{\varphi(r+r')-\varphi(r)}{r'}$ exists as $r' \to 0$, thereby showing that the derivative $\varphi'(r)$ exists, and that

$$\varphi'(r) = \frac{\varphi'(0)}{1-r^2}.$$

Hence

$$\varphi(r) = \int_0^r \varphi'(s)\,ds = \frac{\varphi'(0)}{2}\log\frac{1+r}{1-r},$$

i.e.

$$d(0,x) = \frac{\varphi'(0)}{2}\log\frac{1+\|x\|}{1-\|x\|} = \varphi'(0)\,\omega(0,\|x\|) = \varphi'(0)\,k_B(0,x)$$

for all $x \in B$. The transitivity of $\mathrm{Aut}(B)$ completes the proof.

Q.E.D.

Remark. Theorem VI.2.6 should be compared with a similar proposition in Appendix A, characterizing the Poincaré metric on Δ. For finite-dimensional domains, it was established by D.A. Eisenman (now Pelles) in [Eisenman, 1, Proposition 2.18, pp.13-16].

§ 3. A linear representation of Aut(B).

We shall describe a linear representation of the group Aut(B).

Consider the Hilbert space $H \oplus \mathbb{C}$, and on it the sesquilinear form

$$A((x_1, \zeta_1), (x_2, \zeta_2)) = \zeta_1 \overline{\zeta_2} - (x_1 | x_2).$$

Any linear map of $H \oplus \mathbb{C}$ into $H \oplus \mathbb{C}$ is expressed by a "matrix"

$$T = \begin{pmatrix} A & \xi \\ \lambda & a \end{pmatrix} \tag{VI.3.1}$$

where A is a linear map of H into H, ξ a vector in H, λ a linear form on H, and a is a complex number.

The linear map T leaves A invariant if, and only if,

$$A(T(x_1, \zeta_1),\ T(x_2, \zeta_2)) = A((x_1, \zeta_1), (x_2, \zeta_2)),$$

i.e.

$$A((Ax_1 + \zeta_1 \xi, \lambda(x_1) + a\zeta_1), (Ax_2 + \zeta_2 \xi, \lambda(x_2) + a\zeta_2)) =$$
$$= A((x_1, \zeta_1), (x_2, \zeta_2)),$$

i.e.

$$(\lambda(x_1) + a\zeta_1)\overline{(\lambda(x_2) + a\zeta_2)} - (Ax_1 + \zeta_1 \xi | Ax_2 + \zeta_2 \xi) = \tag{VI.3.2}$$
$$= \zeta_1 \overline{\zeta_2} - (x_1 | x_2)$$

for all $\zeta_1, \zeta_2 \in \mathbb{C}$, $x_1, x_2 \in H$.

For $x_2 = 0$, we must have

$$\overline{a\zeta_2}(\lambda(x_1) + a\zeta_1) - \overline{\zeta_2}(Ax_1 | \xi) - \zeta_1 \overline{\zeta_2} \|\xi\|^2 = \zeta_1 \overline{\zeta_2} \tag{VI.3.3}$$

for all $x_1 \in H$, $\zeta_1, \zeta_2 \in \mathbb{C}$. Setting $x_1 = 0$, the latter condition becomes

(VI.3.4) $$|a|^2 - \|\xi\|^2 = 1.$$

This implies that $|a|^2 = 1 + \|\xi\|^2 \geqslant 1$.

In view of (VI.3.4), (VI.3.3) becomes

(VI.3.5) $$\bar{a}\lambda(x) = (Ax|\xi) \qquad \text{for all } x \in H.$$

Setting $\zeta_1 = \zeta_2 = 0$ in (VI.3.2), we have

(VI.3.6) $$\lambda(x_1)\overline{\lambda(x_2)} - (Ax_1|Ax_2) = -(x_1|x_2),$$

in particular

$$-|\lambda(x)|^2 + \|Ax\|^2 = \|x\|^2 \qquad \text{for all } x \in H,$$

and therefore, by (VI.3.5) and (VI.3.4),

(VI.3.7) $$(1+\|\xi\|^2)(-\|x\|^2+\|Ax\|^2) = |(Ax|\xi)|^2.$$

Thus, by (VI.3.4) and the Schwarz inequality,

$$(1+\|\xi\|^2)(\|Ax\|^2-\|x\|^2) \leqslant \|Ax\|^2\|\xi\|^2,$$

i.e.

$$\|Ax\|^2 \leqslant (1+\|\xi\|^2)\|x\|^2.$$

Hence A is continuous - and $\|A\| \leqslant (1+\|\xi\|^2)^{1/2}$ - , and by (VI.3.5) also λ is a continuous linear form, so that T is continuous.

Denoting by $A^\star$ the adjoint of A, formula (VI.3.5) reads now

(VI.3.8) $$\lambda(x) = (x|\frac{1}{a}A^\star\xi) \qquad \text{for all } x \in H,$$

while (VI.3.6) becomes

(VI.3.9) $$(A^\star Ax_1|x_2) - \frac{1}{|a^2|}(x_1|A^\star\xi)(A^\star\xi|x_2) = (x_1|x_2).$$

This equality holds for all $x_1, x_2 \in H$, and therefore

(VI.3.10) $$A^{\star}A = \mathrm{Id} + \frac{1}{1+\|\xi\|^2}\,(\cdot\,|A^{\star}\xi)A^{\star}\xi.$$

Viceversa, it is easily checked that given $A \in L(H)$, $\xi \in H$ and $a \in \mathbb{C}$, satisfying (VI.3.4) and (VI.3.10), the map T defined by (VI.3.1) in terms of $A, \mathcal{A}, \xi$ and of the linear form given by (VI.3.8) leaves the sesquilinear form $\mathcal{A}$ invariant. That proves

Lemma VI.3.1. *Every linear map* $T: \mathbb{C} \oplus H \to \mathbb{C} \oplus H$ *leaving the sesquilinear form* $\mathcal{A}$ *invariant is given by the matrix*

(VI.3.11) $$T = \begin{pmatrix} A & \xi \\ (\cdot\,|\frac{1}{a}A^{\star}\xi) & a \end{pmatrix},$$

where $A \in L(H)$, $\xi \in H$, $a \in \mathbb{C}$ *satisfy* (VI.3.4) *and* (VI.3.10).

Remark. If $A \in L(H)$ satisfies (VI.3.10), then $\|x\| \leqslant \|Ax\|$ for all $x \in H$. Hence A is closed.

Consider now the Hilbert space $H \oplus \mathbb{C}$ with the scalar product $((x_1,\zeta_1)|(x_2,\zeta_2)) = (x_1|x_2) + \zeta_1\overline{\zeta_2}$. Let T be the operator expressed by (VI.3.11), with $A \in L(H)$, $\xi \in H$, $a \in \mathbb{C}$, satisfying condition (VI.3.4). For $(x,\zeta),(y,\sigma) \in H \oplus \mathbb{C}$,

(VI.3.12) $$\begin{aligned}(T(x,\zeta)|(y,\sigma)) &= ((Ax+\zeta\xi,\ (x|\tfrac{1}{a}A^{\star}\xi)+a\zeta)|(y,\sigma)) = \\ &= (x|A^{\star}y) + \zeta(\xi|y) + \bar{\sigma}((x|\tfrac{1}{a}A^{\star}\xi)+a\zeta) = \\ &= (x|A^{\star}(y+\tfrac{\sigma}{a}\xi)) + \zeta((\xi|y)+a\bar{\sigma}).\end{aligned}$$

Hence $(y,\sigma) \perp T(H \oplus \mathbb{C})$ if, and only if,

(VI.3.13) $$A^{\star}(y + \frac{\sigma}{a}\xi) = 0$$

and

(VI.3.14) $$(\xi | y) + a\bar{\sigma} = 0.$$

If A has a dense image, then its adjoint $A^\star$ is injective, and therefore

(VI.3.15) $$y = -\frac{\sigma}{a}\xi .$$

Thus, by (VI.3.14)

$$\frac{\bar{\sigma}}{\bar{a}}(|a|^2 - \|\xi\|^2) = 0.$$

Since a and ξ satisfy (VI.3.4), then we must have $\sigma = 0$, and, by (VI.3.15) also $y = 0$, i.e., T has a dense image.

By (VI.3.12), the adjoint operator $T^\star$ of T is given by the matrix

$$T^\star = \begin{pmatrix} A^\star & \frac{1}{a}A^\star\xi \\ (\cdot|\xi) & \bar{a} \end{pmatrix}$$

We prove now that, if $T^\star$ is injective, then $A^\star$ is injective. Let $y \in H$ be such that $A^\star y = 0$. Let (x,ζ) be such that

$$y = x + \frac{\zeta}{a}\xi , \qquad (x|\xi) + \bar{a}\zeta = 0.$$

Then, by (VI.3.4),

$$(y|\xi) = (x|\xi) + \frac{\zeta}{a}\|\xi\|^2 =$$

$$= -\bar{a}\zeta + \frac{\zeta}{a}\|\xi\|^2 = \frac{\zeta}{a}(\|\xi\|^2 - |a|^2) = -\frac{\zeta}{a} ,$$

i.e.,

$$x = y + (y|\xi)\xi.$$

Since

$$T^{\star}\begin{pmatrix} x \\ \zeta \end{pmatrix} = \begin{pmatrix} A^{\star}(x + \frac{\zeta}{a}\xi) \\ \\ (x|\xi) + \bar{a}\zeta \end{pmatrix} = \begin{pmatrix} A^{\star}y \\ \\ 0 \end{pmatrix} = \begin{pmatrix} 0 \\ 0 \end{pmatrix},$$

if $T^{\star}$ is injective, then $x = 0$, and therefore $\zeta = 0$ and $y=0$. In conclusion, we have proved

<u>Lemma VI.3.2</u>. *Let* $T \in L(H \oplus \mathbb{C})$ *be expressed by* (VI.3.11) *with* $A \in L(H)$, $\xi \in H$, $a \in \mathbb{C}$ *satisfying* (VI.3.4). *Then* T *has a dense image if, and only if,* A *has a dense image.*

Suppose now that A has a dense image. By the Remark following Lemma VI.3.1, A is surjective. We shall prove that T is surjective, i.e., we shall show that, given $(y,\sigma) \in H \oplus \mathbb{C}$, there exist $(x,\zeta) \in H \oplus \mathbb{C}$ for which

$$\begin{cases} Ax + \zeta\xi = y \\ (x|\frac{1}{a} A^{\star}\xi) + a = \sigma. \end{cases}$$

This system yields

$$\begin{cases} (Ax|\xi) + \zeta\|\xi\|^2 = (y|\xi) \\ (Ax|\xi) + \zeta|a|^2 = \sigma\bar{a} \end{cases}$$

which, by (VI.3.4), is equivalent to

$$(Ax|\xi) = (y|\xi)|a|^2 - \sigma\bar{a}\|\xi\|^2, \tag{VI.3.16}$$

$$\zeta = \sigma\bar{a} - (y|\xi). \tag{VI.3.17}$$

Since A is surjective, there is some $x \in H$ for which

$$Ax = y - (\sigma\bar{a} - (y|\xi))\xi.$$

For such x we have the equation

$$(Ax|\xi) = (y|\xi) - (\sigma\bar{a} - (y|\xi))\|\xi\|^2$$

which, by (VI.3.4), coincides with (VI.3.16). That proves that, if A has a dense image, T is surjective.

By (VI.3.17), if $y=0$ and $\sigma=0$, then $\zeta=0$, so that we have to solve the equation $Ax=0$. On the other hand, by (VI.3.10), A is injective. In conclusion we have proved

Theorem VI.3.3. *Let* G' *be the semigroup consisting of all linear maps of* $H \oplus \mathbb{C}$ *into* $H \oplus \mathbb{C}$ *leaving the sesquilinear hermitian form* A *invariant. If* $T \in G'$, *then* T *is continuous and characterized by the matrix* (VI.3.11) *where* $A \in L(H)$, $\xi \in H$, $a \in \mathbb{C}$, *and* A, ξ, a *satisfy* (VI.3.4) *and* (VI.3.10).

Let G *be the group consisting of all invertible elements in* G'. *For any* $T \in G'$ *the following conditions are equivalent:*

a) $T \in G$;

b) T *is a continuous linear bijective operator of* $H \oplus \mathbb{C}$;

c) T *has a dense image;*

d) A *is a continuous linear bijective operator of* H;

e) A *has a dense image.*

Now, let $T_1, T_2 \in G'$ with

$$T_j = \begin{pmatrix} A_j & \xi_j \\ (\cdot \,|\, \frac{1}{a_j} A_j^{\star}\xi_j) & a_j \end{pmatrix} , \tag{VI.3.18}$$

where $A_j \in L(H)$, $\xi_j \in H$, $a_j \in \mathbb{C}$ satisfy (VI.3.4) and (VI.3.10), and let $T = T_1 \circ T_2$. If T is expressed by (VI.3.11), then one checks easily that

$$A = A_1 \circ A_2 + (A_2 \cdot \,|\, \frac{1}{a_2} \xi_2)\xi_1 ,$$

$$\xi = A_1 \xi_2 + a_2 \xi_1 ,$$

$$a = (A_1 \xi_2 | \frac{1}{a_1} \xi_1) + a_1 a_2 .$$

We shall determine the center Z of G'. Let $T_0 = \begin{pmatrix} A_0 & \xi_0 \\ (\cdot | \frac{1}{a_0} A_0^{\star} \xi_0) & a_0 \end{pmatrix}$ be a central element of G. Then for all

$$T = \begin{pmatrix} A & 0 \\ 0 & a \end{pmatrix} \in G$$

we must have $T \circ T_0 = T_0 \circ T$. By (VI.3.10) $A^{\star} A = Id$. Thus, by condition d) in Theorem VI.3.3, the operator A is unitary. Furthermore, $|a| = 1$, by (VI.3.4). Viceversa, if A is unitary and $|a| = 1$, then $T = \begin{pmatrix} A & 0 \\ 0 & a \end{pmatrix} \in G$. Hence we must have in particular $A \circ A_0 = A_0 \circ A$ and this must hold for all unitary operators A of H. By the Schur lemma A_0 is a scalar multiple of the identity:

$$A_0 = \sigma_0 Id , \qquad \sigma_0 \in \mathbb{C}.$$

Condition (VI.3.10) reads then

$$(|\sigma_0|^2 - 1) Id = \frac{|\sigma_0|^2}{1+\|\xi_0\|^2} (\cdot | \xi_0) \xi_0 ,$$

and thereby, being $\dim_{\mathbb{C}} H > 1$, yields $|\sigma_0| = 1$ and $\xi_0 = 0$. Thus, by (VI.3.4), $|a_0| = 1$. For any $T \in G$, given by (VI.3.11), and all $(x,\zeta) \in H \oplus \mathbb{C}$,

$$T \circ T_0 \begin{pmatrix} x \\ \zeta \end{pmatrix} = \begin{pmatrix} A & \xi \\ (\cdot | \frac{1}{a} A^{\star} \xi) & a \end{pmatrix} \begin{pmatrix} \sigma_0 Id & 0 \\ 0 & a_0 \end{pmatrix} \begin{pmatrix} x \\ \zeta \end{pmatrix} =$$

$$= \begin{pmatrix} \sigma_0 Ax + a_0 \zeta \xi \\ \sigma_0 (x | \frac{1}{a} A^{\star} \xi) + a a_0 \zeta \end{pmatrix} ,$$

$$T_0 \circ T \begin{pmatrix} x \\ \zeta \end{pmatrix} = \begin{pmatrix} \sigma_0 Id & 0 \\ 0 & a_0 \end{pmatrix} \begin{pmatrix} A & \xi \\ (\cdot | \frac{1}{a} A^{\star} \xi) & a \end{pmatrix} \begin{pmatrix} x \\ \zeta \end{pmatrix} = \begin{pmatrix} \sigma_0 (Ax + \zeta \xi) \\ a_0 (x | \frac{1}{a} A^{\star} \xi) + a a_0 \zeta \end{pmatrix}.$$

That implies that $a_0 = \sigma_0$, i.e. that $T_0 = \sigma_0 \mathrm{Id}$ with $|\sigma_0| = 1$. In conclusion we have

<u>Lemma VI.3.4</u>. *The center of* G *coincides with the center of* G' *and is given by* $Z = \{e^{i\theta}\mathrm{Id}: \theta \in \mathbb{R}\}$

Let $T = \begin{pmatrix} A & \xi \\ (\cdot | \frac{1}{a} A^\star \xi) & a \end{pmatrix} \in G'$.

By (VI.3.10) and (VI.3.4)

(VI.3.19) $$0 \leqslant (Ax+\xi | Ax+\xi) = (A^\star Ax | x) + (x | A^\star \xi) + (A^\star \xi | x) + \|\xi\|^2 =$$

$$= \|x\|^2 + \frac{1}{1+\|\xi\|^2} |(x|A^\star\xi)|^2 + (x|A^\star\xi) + (A^\star\xi|x) + \|\xi\|^2 =$$

$$= \|x\|^2 - 1 + 1 + \|\xi\|^2 + \frac{1}{1+\|\xi\|^2} |(x|A^\star\xi)|^2 + (x|A^\star\xi) + (A^\star\xi|x) =$$

$$= \|x\|^2 - 1 + \frac{1}{1+\|\xi\|^2} (|(x|A^\star\xi)|^2 + (1+\|\xi\|^2)(x|A^\star\xi) +$$

$$+ (1+\|\xi\|^2)(A^\star\xi|x) + (1+\|\xi\|^2)^2) =$$

$$= \|x\|^2 - 1 + \frac{1}{1+\|\xi\|^2} |(x|A^\star\xi) + 1 + \|\xi\|^2|^2$$

$$= \|x\|^2 - 1 + \frac{1}{|a|^2} |(x|A^\star\xi) + |a|^2|^2 =$$

$$= \|x\|^2 - 1 + |(x|\tfrac{1}{a} A^\star\xi) + a|^2 .$$

Hence, if $x \in B$, then

(VI.3.20) $$(x|\tfrac{1}{a} A^\star\xi) + a \neq 0 .$$

For $x \in B$, let

(VI.3.21) $$x' = \frac{1}{(Ax|\frac{1}{a}\xi) + a} (Ax + \xi) .$$

Then, by (VI.3.19),

$$\|x'\|^2 = \frac{|a|^2}{|(Ax|\xi) + |a|^2|^2} \|Ax+\xi\|^2 =$$

$$= 1 - \frac{|a|^2}{|(Ax|\xi) + |a|^2|^2} (1 - \|x\|^2)$$

so that $x' \in B$. Hence the map $\tilde{T}: x \mapsto x'$ defined by (VI.3.21) is a holomorphic map of B into B.

The construction of x' can be described as follows. First, imbed B as a subset of $H \oplus \{1\}$. For any $(x,1)$, $x \in B$, consider the line

$$\{\tau T \begin{pmatrix} x \\ 1 \end{pmatrix} : \tau \in \mathbb{C}\} = \{\tau \begin{pmatrix} Ax + \xi \\ (x|\frac{1}{a} A^{\star}\xi)+a \end{pmatrix} : \tau \in \mathbb{C}\}.$$

In view of (VI.3.20), there is a unique point belonging to this line and to the affine hyperplane $H \oplus \{1\}$. That is the point $(x',1)$. The above construction implies that the map $T \mapsto \tilde{T}$ is a homeomorphism of the semigroup G' into the semigroup $\mathrm{Hol}(B,B)$.

As a consequence, the restriction of the map $T \mapsto \tilde{T}$ to G is a homeomorphism of the group G into $\mathrm{Aut}(B)$.

We shall prove that the image of G is the entire group $\mathrm{Aut}(B)$.

First of all, by (VI.3.21), $\tilde{T}(0) = \frac{1}{a}\xi$. For any $x_0 \in B\setminus\{0\}$ we solve the equation

$$\frac{1}{(1+\|\xi\|^2)^{1/2}}\,\xi = x_0$$

by taking first $\|\xi\| = \frac{\|x_0\|}{(1-\|x_0\|^2)^{1/2}}$ and then $\xi = \frac{\|\xi\|}{\|x_0\|}\,x_0$. We will find now a symmetric operator $A \in L(H)$ satisfying (VI.3.10) and having a dense image. For a symmetric operator A, (VI.3.10) becomes

$$\text{(VI.3.22)} \qquad A^2x = x + \frac{1}{1+\|\xi\|^2}\,(Ax|\xi)A\xi \qquad \text{for all } x \in H.$$

Suppose that ξ is an eigenvector of A with eigenvalue ρ:

$$A\xi = \rho\xi.$$

Then (VI.3.22) becomes

$$\rho^2 \xi = \xi + \frac{\rho^2 \|\xi\|^2}{1+\|\xi\|^2} \xi = \frac{1+\|\xi\|^2(1+\rho^2)}{1+\|\xi\|^2} \xi ,$$

i.e.

$$\rho^2 = 1 + \|\xi\|^2, \qquad \rho = \pm (1+\|\xi\|^2)^{1/2} .$$

If ρ satisfies this condition, (VI.3.22) reads

$$A^2 x = x + \frac{\rho^2}{1+\|\xi\|^2} (x|\xi)\xi = x + (x|\xi)\xi,$$

and therefore ξ is an eigenvector of A^2 with eigenvalue $1+\|\xi\|^2$, and A^2 is the identity on $\xi^\perp$. Thus the spectrum of A^2 consists of the points 1 and $1+\|\xi\|^2$.

Let A be the positive square root of A^2. Then ξ is an eigenvector of A with eigenvalue $(1+\|\xi\|^2)^{1/2}$, and A is the identity on $\xi^\perp$. In particular A is symmetric and invertible.

That proves that the corresponding map

$$T = \begin{pmatrix} A & \xi \\ (\cdot | \frac{1}{a} A\xi) & a \end{pmatrix}, \qquad a = (1+\|\xi\|^2)^{1/2} ,$$

belongs to G. Since $\widetilde{T}(0) = x_0$, the image of G is transitive on B.

Furthermore $\widetilde{T}(0) = 0$ if, and only if, $\xi = 0$, i.e., by (VI.3.10), $A^\star A = \mathrm{Id}$, $|a| = 1$. If $T \in G$, then, by Theorem VI.3.3, A is surjective, hence unitary. For any $x \in B$

$$\widetilde{T}(x) = \frac{1}{a} Ax$$

where, being $|a| = 1$, $\frac{1}{a} A$ is unitary. Hence, by Lemma VI.1.2, any automorphism of B with a fixed point in 0, belongs to the image of G. That proves that the image of G is $\mathrm{Aut}(B)$. The kernel of the homomorphism $G \to \mathrm{Aut}\, B$ is $\{e^{i\theta}\mathrm{Id}: \theta \in \mathbb{R}\}$, i.e., the center Z of G. Thus we have

Theorem VI.3.5. *The map* $T \to \widetilde{T}$ *defines an isomorphism of* G/Z *onto* Aut(B).

§ 4. Holomorphic isometries and their fixed points.

All holomorphic automorphisms of B are c_B - isometries (i.e. isometries for the Carathéodory distance c_B). If $\dim_{\mathbb{C}} H < \infty$ all holomorphic maps of B into itself which are isometries for c_B are necessarily surjective, and therefore elements of Aut(B). That is no longer the case when H is infinite dimensional.

We will now characterize the image of the semigroup G' by proving

Theorem VI.4.1. *The image of* G' *is the semi-group* Isom(B) *consisting of all holomorphic maps of* B *into* B *which are isometries for the Carathéodory distance* c_B.

Proof. a). We prove first that, for any $T \in G'$,

$$c_B(\widetilde{T}(x), \widetilde{T}(y)) = c_B(x,y) \qquad \text{for all } x,y \in B.$$

Let $T_1, T_2 \in G$ be such that $y = \widetilde{T}_1(0)$, $\widetilde{T}_2 \circ \widetilde{T}(y) = 0$, and let $S = T_2 T T_1 \in G'$.

Let

$$S = \begin{pmatrix} A & \xi \\ (\cdot \,|\, \frac{1}{a} A^{\star} \xi) & a \end{pmatrix}, \tag{VI.4.1}$$

where $A \in L(H)$, $\xi \in H$, $a \in \mathbb{C}$ satisfy (VI.3.4) and (VI.3.10), so that

$$\widetilde{S} = \frac{1}{(A \cdot \,|\, \frac{1}{a}\xi) + a} (A \cdot + \xi) \in \mathrm{Hol}(B,B). \tag{VI.4.2}$$

Since $\tilde{S}(0) = 0$, then $\xi = 0$. Thus, by (VI.3.4) and (VI.3.10), $\tilde{S}$ is a linear isometry of H.

Since $\tilde{T}_1$ and $\tilde{T}_2$ are isometries for c_B, then

$$c_B(\tilde{T}(x),\tilde{T}(y)) = c_B(\tilde{S}(\tilde{T}_1^{-1}(x)),0) =$$

$$= \omega(0, \|\tilde{S}(T_1^{-1}(x))\|) = \omega(0,\|\tilde{T}_1^{-1}(x)\|) =$$

$$= c_B(0, \tilde{T}_1^{-1}(x)) = c_B(\tilde{T}_1(0),x) = c_B(y,x) =$$

$$= c_B(x,y).$$

b). Now let $f \in \mathrm{Hol}(B,B)$ be an isometry for c_B. Let $x,y \in B$, and choose as before $T_1,T_2 \in G$ in such a way that $y = \tilde{T}_1(0)$, $\tilde{T}_2 f(y) = 0$. Let $h = \tilde{T}_2 \circ f \circ \tilde{T}_1$. The map $h \in \mathrm{Hol}(B,B)$ is an isometry for c_B. Since $h(0) = 0$, then

$$c_B(0,h(z)) = c_B(0,z) \qquad \text{for all } z \in B,$$

i.e.

$$\|h(z)\| = \|z\| \qquad \text{for all } z \in B.$$

By Proposition VI.1.8, h is the restriction to B of a linear isometry A of H. Hence $h = \tilde{T}$ where $T \in G'$ is given by the matrix

$$T = \begin{pmatrix} A & 0 \\ 0 & 1 \end{pmatrix}.$$

In conclusion

$$f = \tilde{T}_2^{-1} \circ \tilde{T} \circ \tilde{T}_1^{-1} = T_2^{-1} T \, T_1^{-1}.$$

Q.E.D.

<u>Corollary VI.4.2</u>. *Let* $f \in \mathrm{Hol}(B,B)$ *be an isometry for* c_B. *Then there exist* $A \in L(H)$, $\xi \in H$, $a \in \mathbb{C}$, *satisfying* (VI.3.4) *and* (VI.3.10) *such that*

$$f(x) = \frac{1}{(x|\frac{1}{a} A^{\star}\xi)+a} (Ax + \xi)$$

for all $x \in B$.

Let $S \in G'$ be expressed by (VI.4.1) and let $\widetilde{S}$ be the associated c_B-isometry expressed by (VI.4.2).

Since the denominator in (VI.4.2) does not vanish on B, then we have

<u>Proposition VI.4.3</u>. *For every holomorphic* c_B*-isometry* f *of* B *there exists a continuous injective linear map* h *of* $\mathbb{C} \oplus H$ *into* $\mathbb{C} \oplus H$ *such that, for every* $x \in B$, *the image of the complex line through* $(1,x)$ *is the complex line through* $(1,f(x))$. *In other words, for any* $x \in B$, *there is a* $\sigma \in \mathbb{C}\setminus\{0\}$ *such that*

$$h(1,x) = \sigma(1,f(x)).$$

As in the case of $\text{Aut}(B)$ this proposition implies

<u>Corollary VI.4.4</u>. *Let* f *be a holomorphic* c_B*-isometry of* B, *and let* L *be an affine subspace of* H. *If* $L \cap B \neq \emptyset$, *then* $f(L \cap B)$ *is the intersection of* B *with an affine subspace* L_1 *of* H. *If* L *is closed, then* L_1 *is closed.*

Let $f \in \text{Hol}(B,B)$ be an isometry for the Carathéodory distance c_B of B. According to Theorem VI.4.1, there exist $A \in L(H)$, $a \in \mathbb{C}$, $\xi \in H$ satisfying (VI.3.4) and (VI.3.10), such that

$$f(x) = \frac{a}{1+\|\xi\|^2+(Ax|\xi)}(Ax + \xi) \qquad \text{for all } x \in B.$$

Let $x \in H$. By the Schwarz inequality and by (VI.3.9),

$$|(Ax|\xi)|^2 \leqslant \|\xi\|^2 (\|x\|^2 + \frac{1}{1+\|\xi\|^2} |(Ax|\xi)|^2)$$

for all $x \in H$, i.e.,

$$(1 - \frac{\|\xi\|^2}{1+\|\xi\|^2})|(Ax|\xi)|^2 \leqslant \|\xi\|^2\|x\|^2.$$

Thus

$$\frac{1}{1+\|\xi\|^2}|(Ax|\xi)|^2 \leqslant \|\xi\|^2\|x\|^2 \qquad \text{for all } x \in H.$$

This shows that the function

$$x \mapsto \frac{a}{1+\|\xi\|^2+(Ax|\xi)}$$

is holomorphic on the ball $B(0,((1+\|\xi\|^2)^{1/2})/\|\xi\|)$, and therefore f is the restriction to B of a holomorphic map of $B(0,((1+\|\xi\|^2)^{1/2})/\|\xi\|)$ into H. Denoting by the same symbol f the latter map, we will prove now that f is continuous on $\bar{B}$ for the weak topology.

Let $x_0 \in \bar{B}$ and let $w_0 \in H$. Then

$$(f(x)-f(x_0)|w_0) = a\{\frac{1}{1+\|\xi\|^2+(Ax|\xi)}((Ax|w_0) - (Ax_0|w_0)) +$$

$$+ (\frac{1}{1+\|\xi\|^2+(Ax|\xi)} - \frac{1}{1+\|\xi\|^2+(Ax_0|\xi)})(Ax_0|w_0) +$$

$$+ (\frac{1}{1+\|\xi\|^2+(Ax|\xi)} - \frac{1}{1+\|\xi\|^2+(Ax_0|\xi)})(\xi|w_0)\}.$$

Let $\varepsilon > 0$. Choosing $w_1 = A^{\star}\xi$, for any $\sigma > 0$ there is a $\delta_1 > 0$ such that, whenever $x \in \bar{B}$ satisfies the condition

$$|(x|w_1) - (x_0|w_1)| < \delta_1 ,$$

then

$$\left|\frac{1}{1+\|\xi\|^2+(Ax|\xi)} - \frac{1}{1+\|\xi\|^2+(Ax_0|\xi)}\right| < \sigma ,$$

and therefore also

$$\left|\frac{1}{1+\|\xi\|^2+(Ax|\xi)}\right| \leqslant \left|\frac{1}{1+\|\xi\|^2+(Ax_0|\xi)}\right| + \sigma.$$

Choosing $w_2 = A^{\star}w_0$, and setting $\delta = \inf(\sigma,\delta_1)$, whenever $x \in \bar{B}$ is such that

$$|(x|w_j) - (x_0|w_j)| < \delta \qquad (j=1,2),$$

then

$$|(f(x)-f(x_0)|w_0)| < \sigma|a|\{(\left|\frac{1}{1+\|\xi\|^2+(Ax_0|\xi)}\right|+\sigma)+|(x_0|w_2)+(\xi|w_0)|\}.$$

Choosing $\sigma > 0$ so small that the right hand side be smaller than ε, we conclude with

Theorem VI.4.5. *Every holomorphic isometry of* B *into* B *for the Carathéodory distance is the restriction to* B *of a holomorphic map of a ball* B(0,r), *with radius* $r > 1$, *into* H. *The restriction of this map to* $\overline{B}$ *is continuous for the weak topology.*

By the Banach-Alaoglu theorem, the convex set $\overline{B}$ is compact for the weak topology. Thus the Schauder-Tychonoff fixed point theorem [Dunford-Schwartz, 1, p.456] yields

Theorem VI.4.6. *Every holomorphic isometry of* B *for the Carathéodory distance (is the restriction to* B *of a holomorphic map of a neighborhood of* $\overline{B}$ *in* H, *which) has a fixed point in* $\overline{B}$.

In particular, every holomorphic automorphism of B has a fixed point in $\overline{B}$ [Hayden-Suffridge, 1].

Theorem VI.4.7. *Let* f *be a holomorphic* c_B*-isometry of* B. *The set of fixed points of* f *in* B, *if not empty, is the intersection of* B *with a closed affine subspace of* H.

Proof. Let $x_0 \in B$ be a fixed point of f and let $g \in \mathrm{Aut}(B)$ be such that $g(x_0) = 0$. Then 0 is the fixed point of the c_B-isometry $h = g \circ f \circ g^{-1}$, and therefore h is linear. Hence the set F of fixed points of h in H is a closed vector subspace of H. The set of fixed points of f in B is $g^{-1}(B \cap F)$, which, by Theorem VI.1.7, is the intersection of B with a closed affine subspace of H.

Q.E.D.

The following theorem, due to T.L. Hayden and T.J. Suffridge, describes the set of fixed points on the boundary in case there are no fixed points in B.

Theorem VI.4.8. *If* $g \in \operatorname{Aut} B$ *has no fixed point in* B, *then its fixed point set in* $\overline{B}$ *consists of one or two points.*

Proof. By Theorem VI.4.6, g has at least one fixed point on the boundary ∂B of B. If there are two fixed points, g leaves invariant the affine line L joining them. Since all points of ∂B are extreme points of $\overline{B}$, then $L \cap B \neq \emptyset$. Let $f \in \operatorname{Aut}(B)$ be such that $0 \in f(L \cap B)$. Then $f(L \cap B)$ is the intersection of B with a complex line, which is invariant under the action of the automorphism $h = f \circ g \circ f^{-1}$. The restriction of h to $f(L \cap B)$ is a Moebius transformation. Let v be one of the two fixed points of h in $\overline{f(L \cap B)}$. Then $\|v\| = 1$. Setting $b = h^{-1}(0)$, then $b \in f(L \cap B)$, and therefore $b = \zeta_0 v$ with $\zeta_0 = (b|v)$, and $|\zeta_0| < 1$.

By Theorem VI.1.3, $h = U \circ f_b$, where U is a unitary operator of H and f_b is expressed by (VI.1.6). Since

$$f_b(v) = \frac{1-\zeta_0}{1-\overline{\zeta}_0}\, v,$$

then

$$U(v) = \frac{1-\overline{\zeta}_0}{1-\zeta_0}\, v,$$

i.e., v is an eigenvector of U, with eigenvalue $\dfrac{1-\overline{\zeta}_0}{1-\zeta_0}$.

Thus, by (VI.1.6), we have, for all $x \in \overline{B}$,

$$h(x) = U \circ f_b(x) = \frac{(x|v)-\zeta_0}{1-\overline{\zeta}_0 (x|v)}\, \frac{1-\overline{\zeta}_0}{1-\zeta_0}\, v + \frac{\alpha(\zeta_0)}{1-\overline{\zeta}_0 (x|v)}\, U(x-(x|v)v).$$

Since $x-(x|v)v \perp v$ and U is unitary, then

$$U(x-(x|v)v \perp v. \tag{VI.4.3}$$

The fixed points of h are the images, by f, of the fixed points of g. Hence let w be a fixed point of h, distinct from v.

Then

$$w = \frac{(w|v)-\zeta_0}{1-\bar{\zeta}_0 (w|v)} \frac{1-\bar{\zeta}_0}{1-\zeta_0} v + \frac{\alpha(\zeta_0)}{1-\bar{\zeta}_0 (w|v)} U(w-(w|v)v),$$

and therefore

$$(w|v) = \frac{(w|v)-\zeta_0}{1-\bar{\zeta}_0 (w|v)} \frac{1-\bar{\zeta}_0}{1-\zeta_0} . \tag{VI.4.4}$$

Consider the Moebius transformation

$$\zeta \mapsto \frac{1-\bar{\zeta}_0}{1-\zeta_0} \frac{\zeta-\zeta_0}{1-\bar{\zeta}_0 \zeta} .$$

One of its fixed points is $\zeta = 1$. Thus all its fixed points must have modulus one. Hence, by (VI.4.4), $|(w|v)| = 1$, i.e., $w = e^{i\theta}v$, for some $\theta \in \mathbb{R}$, and therefore w is the other fixed point of h on $f(L \cap B)$. Hence h has no fixed points on $\bar{B}$ except v and w. In conclusion g has no fixed points on $\bar{B}$ except $f(v)$ and $f(w)$.

Q.E.D.

Now, let f be a holomorphic c_B-isometry of B with no fixed point in B. By Theorem VI.4.6, f has at least one fixed point on the boundary ∂B of B. If there are two fixed points, the affine complex line joining them intersects B on a disc which is invariant under f. The restriction of f to the disc is a Moebius transformation with no fixed point inside the disc. Thus the restriction has only two fixed points on the boundary of the disc. This shows that, if the set of fixed points contains more than two points, no more than two of them lie on the same complex affine line. Thus three fixed points determine a unique two-dimensional complex affine subspace L of $\mathcal{H}$; L is closed and $L \cap B$ is invariant by f. By Lemma

VI.2.3, the restriction $f_{|L \cap B}$ is a holomorphic $c_{L \cap B}$-isometry of $L \cap B$. The affine space L being finite dimensional, $f_{|L \cap B}$ is an automorphism of the ball $L \cap B$. Since $f_{|L \cap B}$ has no fixed point inside $L \cap B$, by Theorem VI.4.8, $f_{|L \cap B}$ can have only two fixed points on the boundary. This contradiction proves that the conclusion of Theorem VI.4.8 holds also for c_B-isometries, i.e.,

Theorem VI.4.9. *Let* f *be a holomorphic* c_B*-isometry of* B. *If* f *has no fixed point in* B, *then* f *has at most two fixed points on the boundary of* B.

By Brouwer's classical result, if H has finite dimension, every continuous map $\overline{B} \to \overline{B}$ has a fixed point. The following example of S. Kakutani [Kakutani, 1] shows that holomorphy is a crucial condition in the infinite dimensional case.

Theorem VI.4.10. *Let* H *be a real separable, infinite dimensional Hilbert space and let* B *be the open unit ball of* H. *There exists a homeomorphism of* $\overline{B}$ *onto* $\overline{B}$ *with no fixed point.*

Proof. Let $(e_n)_{n \in \mathbb{Z}}$ be a complete orthonormal system in H and let U be the right shift operator on H, i.e., the linear operator on H defined by

$$U(e_n) = e_{n+1} \qquad n \in \mathbb{Z}.$$

We recall a few facts about U.

For all $x = \Sigma\, a_n e_n \in H$, $y = \Sigma\, b_n e_n \in H$, we have $(Ux, Uy) = (x, y)$. Thus U, being surjective, is a unitary operator in H. Let $x \in H$ and $\theta \in \mathbb{R}$ be such that $U(x) = e^{i\theta}x$. That is equivalent to the equation

$$\Sigma\, a_{n-1}\, e_n = \Sigma\, e^{i\theta} a_n\, e_n ,$$

i.e.,

$$a_{n-1} = e^{i\theta} a_n \qquad \text{for all} \quad n \in \mathbb{Z} .$$

Since $\|x\|^2 = \Sigma\, |a_n|^2 < \infty$, that implies that $a_n = 0$ for all $n \in \mathbb{Z}$, i.e., $x = 0$. Thus U has no eigenvalues. In particular there is no vector $x \neq 0$ such that $U(x) = x$. That is to say that U *has no fixed point on* $H\setminus\{0\}$.

Let B be the open unit ball of H, and let $F\colon \overline{B} \to H$ be the continuous map defined by

$$F(x) = \frac{1-\|x\|}{2}\, e_0 + U(x).$$

For $x \in B\setminus\{0\}$, let $y = \frac{1}{\|x\|}\, x$. Then

$$F(x) = \frac{1}{2}\,(1-\|x\|)e_0 + \|x_0\| U(y).$$

This shows that $F(x)$ divides the segment $[\frac{1}{2}\, e_0 , U(y)]$ in two parts which are proportional to $\|x\|$ and $1-\|x\|$. Since $F(0) = \frac{1}{2}\, e_0$, we conclude that F is a homeomorphism of $\overline{B}$ onto $\overline{B}$.

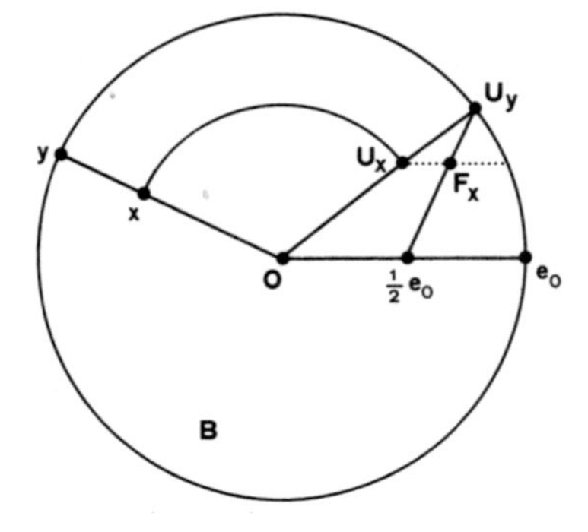

We prove now that F has no fixed point in $\overline{B}$.

Suppose that $F(x_0) = x_0$, for some $x_0 \in \overline{B}$. Then

$$x_0 - U(x_0) = \frac{1-\|x_0\|}{2}\, e_0 . \tag{VI.4.5}$$

This shows first of all that $x_0 \neq 0$. Since the restriction of F to the unit sphere $S = \{\|x\| = 1\}$ is U, and U has no fixed point on $H \setminus \{0\}$, then we must have $\|x_0\| < 1$.

Let $x_0 = \Sigma\, a_n e_n$, with $0 < \Sigma |a_n|^2 = \|x_0\|^2 < 1$. Then

$$x_0 - U(x_0) = \Sigma\, (a_n - a_{n-1}) e_n .$$

Thus, by (VI.4.5), we must have

$$\dots = a_{-2} = a_{-1} < a_0 = a_1 = a_2 = \dots .$$

Since $\Sigma |a_n|^2 < \infty$, then we must have $a_n = 0$ for all $n \in \mathbb{Z}$, i.e., $x_0 = 0$. This contradiction proves Theorem VI.4.10.

Q.E.D.

Notes.

The construction of the group of all holomorphic automorphisms of the unit ball B in a complex Hilbert space H can be found in [Hervé, 2] for the finite-dimensional case (cf. also, e.g., [Siegel, 2 and 3] and [Hua, 1]) and in [Hayden-Suffridge, 1] and [Renaud, 1] for infinite dimensional spaces.

The results on the fixed points of holomorphis automorphism were established by T.L. Hayden and T.J. Suffridge in [Hayden-Suffridge, 1]. Their argument consists of showing that any holomorphic automorphism of B extends to a holomorphic map of a neighborhood of the closure of B and then of proving that the extension is weakly continuous. Holomorphic extendibility is a general property of all holomorphic automorphisms of the unit ball of any complex Banach space, as it was shown by W. Kaup and H. Upmeier in [Kaup-Upmeier, 1]. However, the example of J.L. Stachò given in § 4 of chapter IV shows that the weak continuity of the holomorphic extension is not a general property.

The linear representation of Aut(B) described in § 3 extends to the infinite dimensional case the classical relationship between the group U(n,1) and the group Aut(B) when B is the open euclidean unit ball in $\mathbb{C}^n$. (Cf. e.g. [Siegel, 2 and 3], [Hua, 1]). The construction, for the unit ball $B \in H$, of the Carathéodory distance (whose topological characterization in § 2, extends the one given by D.A. Eisenman (now D.A. Pelles) in [Eisenman, 1]) leads naturally to the semigroup of all holomorphic isometries of B and to the linear representation discussed in § 4.

The example of Kakutani discussed in § 4 is in [Kakutani, 1].

APPENDIX A

THE POINCARÉ METRIC

I. Let $\Delta = \{\zeta \in \mathbb{C}: |\zeta| < 1\}$ be the unit disc in $\mathbb{C}$ and let $f \in \mathrm{Hol}(\Delta)$ be a holomorphic function on Δ such that $f(\Delta) \subset \subset \overline{\Delta}$ (the closure of Δ). Setting $\partial\Delta = \{\zeta \in \mathbb{C}: |\zeta| = 1\}$, the maximum principle implies that, if $f(\Delta) \cap \partial\Delta \neq \emptyset$, then f is constant: $f(\zeta) = \zeta_0$ for all $\zeta \in \Delta$, where $\zeta_0 \in \partial\Delta$.

<u>Theorem A.1</u> (Schwarz lemma). *Let* $f(0) = 0$. *Then*

$$|f(\zeta)| \leqslant |\zeta| \quad \textit{for all} \quad \zeta \in \Delta \tag{A.1}$$

and

$$|f'(0)| \leqslant 1.$$

If $|f(\zeta_0)| = |\zeta_0|$ *for some* $\zeta_0 \in \Delta\setminus\{0\}$, *or if* $|f'(0)| = 1$, *then there exists* $\theta \in \mathbb{R}$ *such that* $f(\zeta) = e^{i\theta}\zeta$.

<u>Proof.</u> Let

$$f(\zeta) = a_1\zeta + a_2\zeta^2 + \dots \qquad (a_j \in \mathbb{C}, \quad j = 1,2,\dots)$$

be the power series expansion of f in Δ. The function $\zeta \mapsto \frac{f(\zeta)}{\zeta} = a_1 + a_2\zeta + \dots$ is holomorphic on Δ, and for all $0 < r < 1$ and $|\zeta| = r$, we have

$$\left|\frac{f(\zeta)}{\zeta}\right| = \frac{|f(\zeta)|}{r} \leqslant \frac{1}{r}.$$

By the maximum principle, the above inequality holds also for $|\zeta| \leqslant r$. Letting $r \nearrow 1$ we obtain (A.1). If $|f(\zeta_0)| = |\zeta_0|$ for some $\zeta_0 \in \Delta\setminus\{0\}$, the function $\zeta \mapsto \left|\frac{f(\zeta)}{\zeta}\right|$ reaches its maximum, 1, at the point ζ_0 and therefore, by the maximum principle, the holomorphic function $\zeta \mapsto \frac{f(\zeta)}{\zeta}$ is a constant of modulus one.

Being $f'(0) = \left(\frac{f(\zeta)}{\zeta}\right)_{\zeta=0} = a_1$, by the maximum principle we have $|a_1| \leqslant 1$, equality holding if, and only if, $f(\zeta) = e^{i\theta}\zeta$ for all $\zeta \in \Delta$ and some $\theta \in \mathbb{R}$.

Let $\mathrm{Aut}(\Delta)$ be the group of all holomorphic automorphisms of Δ.

Corollary A.2. *If* $f \in \mathrm{Aut}(\Delta)$ *and* $f(0) = 0$, *then* f *is a rotation, i.e., there exists* $\theta \in \mathbb{R}$ *such that* $f(\zeta) = e^{i\theta}\zeta$ *for all* $\zeta \in \Delta$.

For $\zeta_0 \in \Delta$ and $\theta \in \mathbb{R}$ let $g \in \mathrm{Hol}(\Delta)$ be the holomorphic function defined by

$$g(\zeta) = e^{i\theta}\,\frac{\zeta-\zeta_0}{1-\bar{\zeta}_0\zeta}\,. \tag{A.2}$$

Since

$$1 - |g(\zeta)|^2 = \frac{|\zeta_0|^2}{|1-\bar{\zeta}_0\zeta|^2}\,(1-|\zeta|^2) > 0 \quad \text{for all} \quad \zeta \in \Delta,$$

then $g(\Delta) \subset \Delta$. Furthermore

$$\zeta = e^{-i\theta}\,\frac{g(\zeta) + e^{i\theta}\zeta_0}{1 + \overline{e^{i\theta}\zeta_0}\,g(\zeta)} \qquad (\zeta \in \Delta).$$

Hence $g \in \mathrm{Aut}(\Delta)$; g is called a *Moebius transformation*.

For $\zeta_1, \zeta_2 \in \Delta$, $\theta_1, \theta_2 \in \mathbb{R}$, let g_1, g_2 be the Moebius transformations defined by

$$g_1(\zeta) = e^{i\theta_1}\,\frac{\zeta-\zeta_1}{1-\bar{\zeta}_1\zeta}\,, \quad g_2(\zeta) = e^{i\theta_2}\,\frac{\zeta-\zeta_2}{1-\bar{\zeta}_2\zeta} \qquad (\zeta \in \Delta).$$

For all $\zeta \in \Delta$

$$g_1(g_2(\zeta)) = e^{i(\theta_1+\theta_2)} \frac{\zeta - h(\zeta)}{1-\overline{h(\zeta_1)}\zeta},$$

where h is the Moebius transformation

$$h(\zeta) = e^{-i\theta_2} \frac{\zeta + e^{i\theta_2}\zeta_2}{1+\overline{e^{i\theta_2}\zeta_2}\zeta}.$$

Hence the family of all Moebius transformations is a group. Since this group is obviously transitive on Δ, for any $f \in \in \mathrm{Aut}(\Delta)$ let g be the Moebius transformation (A.2) with $\zeta_0 = f(0)$, $\theta = 0$.

Being $g(f(0)) = 0$, by Corollary A.2, there exists $\theta \in \in \mathbb{R}$ such that

$$g(f(\zeta)) = e^{i\theta}\zeta ,$$

i.e.,

$$f(\zeta) = g^{-1}(e^{i\theta}\zeta) \qquad \text{for all } \zeta \in \Delta.$$

That proves

<u>Theorem A.3</u>. *The group* $\mathrm{Aut}(\Delta)$ *is the group of all Moebius transformations.*

As a consequence of (A.2) and of Theorem A.3, all holomorphic automorphisms of Δ are restrictions to Δ of holomorphic functions defined on neighborhoods of $\bar{\Delta}$.

We will now give a different description of $\mathrm{Aut}(\Delta)$.

The group $U(1,1)$ is, by definition, the group of all linear operators of $\mathbb{C}^2$ leaving the hermitian form $|\zeta_1|^2 - |\zeta_2|^2$ invariant, i.e. the group of all 2×2 complex matrices $\begin{pmatrix} a & b \\ c & d \end{pmatrix}$ such that

$$\overline{{}^t\begin{pmatrix} a & b \\ c & d \end{pmatrix}} \begin{pmatrix} 1 & 0 \\ 0 & -1 \end{pmatrix} \begin{pmatrix} a & b \\ c & d \end{pmatrix} = \begin{pmatrix} 1 & 0 \\ 0 & -1 \end{pmatrix}.$$

This condition is equivalent to

$$|a|^2 - |c|^2 = 1, \quad |d|^2 - |b|^2 = 1, \quad \bar{a}b = \bar{c}d,$$

which imply $|a| = |d| > 1$, and may also be written:

$$a = e^{i\alpha} \cosh t, \quad b = e^{i\beta} \sinh t, \quad c = e^{i\gamma} \sinh t, \quad d = e^{i\delta} \cosh t, \tag{A.3}$$

where $t, \alpha, \beta, \gamma, \delta$ are real numbers such that either $t=0$ or

$$\beta - \alpha \equiv \delta - \gamma \qquad (\mathrm{mod}\ 2\pi). \tag{A.4}$$

For all $\begin{pmatrix} a & b \\ c & d \end{pmatrix} \in U(1,1)$, $\left|\det \begin{pmatrix} a & b \\ c & d \end{pmatrix}\right| = 1$. The subgroup $SU(1,1)$ consisting of all the elements in $U(1,1)$ having determinant one is obviously normal. Since by (A.3) and (A.4) $ad - bc = e^{i(\alpha+\delta)}$, then $\begin{pmatrix} a & b \\ c & d \end{pmatrix} \in SU(1,1)$ if, and only if,

$$\alpha + \delta \equiv \beta + \gamma \equiv 0 \qquad (\mathrm{mod}\ 2\pi).$$

Thus $SU(1,1)$ consists of all 2×2 complex matrices $\begin{pmatrix} a & b \\ \bar{b} & \bar{a} \end{pmatrix}$ such that

$$|a|^2 - |b|^2 = 1. \tag{A.5}$$

For any such matrix the function

$$\phi(g): \zeta \mapsto \frac{a\zeta+b}{\bar{b}\zeta+\bar{a}}$$

is holomorphic on (a neighborhood of the closure of) Δ. Being

$$\frac{a\zeta+b}{\bar{b}\zeta+\bar{a}} = \frac{a}{\bar{a}} \frac{\zeta + \frac{b}{a}}{1 + \frac{\bar{b}}{\bar{a}}\zeta} \tag{A.6}$$

and $\left|\frac{a}{\bar{a}}\right| = 1$, $\left|\frac{b}{a}\right| < 1$, then $\phi(g) \in \mathrm{Aut}(\Delta)$. A trivial computation shows that the map $\phi: SU(1,1) \to \mathrm{Aut}(\Delta)$ is a homomorphism, whose kernel is the subgroup $\pm\begin{pmatrix}1 & 0\\ 0 & 1\end{pmatrix}$, which is the center of $SU(1,1)$.

We prove now that ϕ is surjective.

Given any Moebius transformation

$$\zeta \mapsto e^{i\theta}\,\frac{\zeta+\zeta_0}{1+\bar{\zeta}_0\zeta}$$

with $\zeta_0 \in \Delta$, $\theta \in \mathbb{R}$, comparing with (A.6) we have to find a and b satisfying (A.5) and such that $\theta = 2\arg a$, $\frac{b}{a} = \zeta_0$. We choose

$$a = (1 - |\zeta_0|^2)^{-\frac{1}{2}}\, e^{i\frac{\theta}{2}}, \quad b = \zeta_0 a.$$

In conclusion we have proved

<u>Theorem A.4</u>. *The map* ϕ *is a homomorphism of* $SU(1,1)$ *onto* $Aut(\Delta)$, *whose kernel is the center of* $SU(1,1)$.

Finally the map $SU(1,1) \times \Delta \to \Delta$ defined by $(g,\zeta) \mapsto \phi(g)(\zeta)$ is analytic (for the natural Lie group structure on $SU(1,1)$ and the real structure underlying the complex structure on Δ).

II. Going back to the Schwarz lemma, let $f \in Hol(\Delta)$ be such that $f(\Delta) \subset \Delta$. For any $\zeta_0 \in \Delta$ consider the Moebius transformations

$$h_1 : \zeta \mapsto \frac{\zeta+\zeta_0}{1+\bar{\zeta}_0\zeta}, \quad h_2 : \zeta \mapsto \frac{\zeta - f(\zeta_0)}{1-\overline{f(\zeta_0)}\zeta}.$$

Being $(h_2 \circ f \circ h_1)(\Delta) \subset \Delta$, and $(h_2 \circ f \circ h_1)(0) = 0$, the Schwarz lemma yields:

(A.7) $\quad |(h_2 \circ f \circ h_1)(\zeta)| \leqslant |\zeta| \quad$ for all $\zeta \in \Delta$,

(A.8) $\quad |(h_2 \circ f \circ h_1)'(0)| \leqslant 1;$

if $f \in Aut(\Delta)$ both inequalities become equalities; viceversa if either the second one is an equality or if the first one becomes an equality at some $\zeta \in \Delta\setminus\{0\}$, then $f \in \mathrm{Aut}(\Delta)$. Since

$$h_1^{-1}(\zeta) = \frac{\zeta-\zeta_0}{1-\bar{\zeta}_0\zeta},$$

then (A.7) becomes

$$\left|\frac{f(\zeta) - f(\zeta_0)}{1 - \overline{f(\zeta_0)}f(\zeta)}\right| \leqslant \left|\frac{\zeta - \zeta_0}{1 - \bar{\zeta}_0\zeta}\right| \quad \text{for all} \quad \zeta \in \Delta.$$

Being

$$h_1'(0) = 1 - |\zeta_0|^2, \quad h_2'(f(\zeta_0)) = \frac{1}{1 - |f(\zeta_0)|^2},$$

(A.8) becomes

$$\frac{|f'(\zeta_0)|(1 - |\zeta_0|^2)}{1 - |f(\zeta_0)|^2} \leqslant 1.$$

In conclusion the Schwarz lemma can now be re-stated in the following form:

Theorem A.5. (Schwarz-Pick lemma). *For any* $f \in \mathrm{Hol}(\Delta)$ *such that* $f(\Delta) \subset \Delta$ *and for any choice of* $\zeta_0, \zeta_1, \zeta_2$ *in* Δ, *we have*

$$\left|\frac{f(\zeta_1) - f(\zeta_2)}{1 - \overline{f(\zeta_2)}f(\zeta_1)}\right| \leqslant \left|\frac{\zeta_1 - \zeta_2}{1 - \bar{\zeta}_2\zeta_1}\right|,$$

$$\left|\frac{f'(\zeta_0)}{1 - |f(\zeta_0)|^2}\right| \leqslant \frac{1}{1 - |\zeta_0|^2}.$$

If $f \in \mathrm{Aut}(\Delta)$, *both inequalities become equalities. Viceversa, if equality holds for some* $\zeta_0 \in \Delta$ *or for a pair of points* $\zeta_1 \neq \zeta_2$ *in* Δ, *then* $f \in \mathrm{Aut}(\Delta)$.

Let us introduce in Δ the Riemannian metric (Poincaré hyperbolic metric)

$$ds^2 = (1 - |\zeta|^2)^{-2}|d\zeta|^2,$$

and for $v \in \mathbb{C}$, $\zeta \in \Delta$, set

$$\langle v \rangle_\zeta = \frac{|v|}{1 - |\zeta|^2}.$$

The Schwarz-Pick lemma implies

<u>Proposition A.6</u>. *For any* $f \in \mathrm{Hol}(\Delta)$ *such that* $f(\Delta) \subset \Delta$, *the differential* df *contracts the Poincaré hyperbolic metric, i.e.*

$$\langle f'(\zeta) \rangle_{f(\zeta)} \leqslant \langle 1 \rangle_{\zeta} \qquad \textit{for all} \quad \zeta \in \Delta.$$

If $f \in \mathrm{Aut}(\Delta)$ *equality holds for all* $\zeta \in \Delta$. *Viceversa, if equality holds at some* $\zeta \in \Delta$, *then* $f \in \mathrm{Aut}(\Delta)$.

Now, let f be an isometry for the Poincaré hyperbolic metric. Setting $u = \mathrm{Re}\, f$, $v = \mathrm{Im}\, f$, $\xi = \mathrm{Re}\,\zeta$, $\eta = \mathrm{Im}\,\zeta$, then we must have

$$\frac{1}{(1-(u^2+v^2))^2}\left((du)^2+(dv)^2\right) = \frac{1}{(1-(\xi^2+\eta^2))}\left((d\xi)^2+(d\eta)^2\right)$$

and that implies

$$\begin{cases} \left(\dfrac{\partial u}{\partial \xi}\right)^2 + \left(\dfrac{\partial v}{\partial \xi}\right)^2 = \left(\dfrac{\partial u}{\partial \eta}\right)^2 + \left(\dfrac{\partial v}{\partial \eta}\right)^2 \\[2ex] \dfrac{\partial u}{\partial \xi}\dfrac{\partial u}{\partial \eta} + \dfrac{\partial v}{\partial \xi}\dfrac{\partial v}{\partial \eta} = 0 \end{cases}$$

i.e.

$$\frac{\partial u}{\partial \xi} = t\,\frac{\partial v}{\partial \eta}\,, \qquad \frac{\partial u}{\partial \xi} = -t\,\frac{\partial u}{\partial \eta}$$

with $t = \pm 1$. Thus f is a holomorphic or antiholomorphic function of ζ. In conclusion

<u>Proposition A.7</u>. *The group of all isometries for the Poincaré metric consists of all holomorphic and anti-holomorphic automorphisms of* Δ.

The Gaussian curvature of the Poincaré hyperbolic metric is

$$K = -\frac{2}{g}\,\frac{\partial^2 \log g}{\partial\zeta\partial\bar{\zeta}} = -4(1-|\zeta|^2)^2\,\frac{1-|\zeta|^2+\zeta\bar{\zeta}}{(1-|\zeta|^2)^2} = -4.$$

III. Let $\omega : \Delta \times \Delta \to \mathbb{R}^+$ be the distance function defined by the Poincaré hyperbolic metric. For $\zeta_1, \zeta_2 \in \Delta$,

$$\omega(\zeta_1, \zeta_2) = \inf \int_\ell ds,$$

where inf is taken over all rectifiable arcs ℓ in Δ, joining ζ_1 and ζ_2.

For $\zeta_1 = 0$, $\zeta_2 = \xi_2$ with $0 < \xi_2 < 1$,

$$\omega(0, \xi_2) = \int_0^{\xi_2} \frac{d\xi}{1-\xi^2} = \frac{1}{2} \log \frac{1+\xi_2}{1-\xi_2} .$$

Since $\zeta \mapsto e^{i\theta}\zeta$ $(\theta \in \mathbb{R})$ is a (holomorphic) automorphism of Δ, and since the Poincaré metric is invariant under $\operatorname{Aut} \Delta$, then

(A.9) $$\omega(0, \zeta) = \omega(0, |\zeta|) = \frac{1}{2} \log \frac{1+|\zeta|}{1-|\zeta|} .$$

Note that we have shown that the geodesic lines at 0 for the Poincaré hyperbolic metric are the radii of the circle $\partial\Delta$.

For $\zeta_1, \zeta_2 \in \Delta$, the Moebius transformation

$$\zeta \mapsto \frac{\zeta - \zeta_1}{1 - \bar{\zeta}_1 \zeta}$$

maps ζ_1 into 0 and ζ_2 into $\dfrac{\zeta_2 - \zeta_1}{1 - \bar{\zeta}_1 \zeta_2}$. Hence the Poincaré distance $\omega(\zeta_1, \zeta_2)$ is given by

$$\omega(\zeta_1, \zeta_2) = \omega\left(0, \frac{\zeta_2 - \zeta_1}{1 - \bar{\zeta}_1 \zeta_2}\right) = \frac{1}{2} \log \frac{1 + \left|\dfrac{\zeta_2 - \zeta_1}{1 - \bar{\zeta}_1 \zeta_2}\right|}{1 - \left|\dfrac{\zeta_2 - \zeta_1}{1 - \bar{\zeta}_1 \zeta_2}\right|} .$$

Since the function $t \mapsto \log \frac{1+t}{1-t}$ is strictly increasing on the interval $[0,1)$, the Schwarz-Pick lemma yields

<u>Proposition A.8</u>. *Any* $f \in \operatorname{Hol}(\Delta)$, *such that* $f(\Delta) \subset \Delta$, *contracts the Poincaré distance:*

(A.10) $\qquad \omega(f(\zeta_1),f(\zeta_2)) \leqslant \omega(\zeta_1,\zeta_2) \qquad$ *for all* $\zeta_1,\zeta_2 \in \Delta$.

Hence, the holomorphic automorphisms of Δ *are isometries for* ω. *Conversely, if a holomorphic map* $f\colon \Delta \to \Delta$ *is such that equality holds in* (A.10) *at two distinct points* ζ_1,ζ_2 *in* Δ, *then* $f \in \mathrm{Aut}(\Delta)$.

The following proposition characterizes the Poincaré distance.

<u>Proposition A.9</u>. *Let* $\sigma\colon \Delta \times \Delta \to \mathbb{R}$ *be a function satisfying the following three conditions:*

1) σ *is invariant under* $\mathrm{Aut}(\Delta)$, *i.e.,*

$$\sigma(f(\zeta_1),f(\zeta_2)) = \sigma(\zeta_1,\zeta_2)$$

for all $\zeta_1,\zeta_2 \in \Delta$, $f \in \mathrm{Aut}(\Delta)$.

2) For real ζ_1,ζ_2, *such that* $0<\zeta_1<\zeta_2<1$,

$$\sigma(0,\zeta_2) = \sigma(0,\zeta_1)+\sigma(\zeta_1,\zeta_2);$$

3) $\displaystyle\lim_{t\searrow 0} \frac{\sigma(0,t)}{t} = 1.$

Then

$$\sigma(\zeta_1,\zeta_2) = \omega(\zeta_1,\zeta_2)$$

for all $\zeta_1\ \zeta_2 \in \Delta$.

<u>Proof</u>. Let r and δ be real numbers such that

$$0 < r < r+\delta < 1.$$

By 2)

$$\sigma(0,\ r+\delta) = \sigma(0,r) + \sigma(r,\ r+\delta),$$

and by 1)

$$\sigma(r,\ r+\delta) = \sigma(0,\ \frac{r+\delta-r}{1-r(r+\delta)}) = \sigma(0,\ \frac{\delta}{1-r\delta-r^2}) .$$

Thus

$$\frac{1}{\delta}(\sigma(0,r+\delta) - \sigma(0,r)) = \frac{1}{\delta}\sigma(0,\frac{\delta}{1-r\delta-r^2}) = \frac{1}{1-r\delta-r^2}\ \frac{1-r\delta-r^2}{\delta}\sigma(0,\frac{\delta}{1-r\delta-r^2}) .$$

Letting $\delta \searrow 0$ and applying 3) we see that

$$\lim_{\delta \searrow 0} \frac{1}{\delta}(\sigma(0,r+\delta) - \sigma(0,r)) = \frac{1}{1-r^2} .$$

Now, let r and δ be real numbers such that

$$0 < r+\delta < r < 1.$$

Again, by 2)

$$\sigma(0,r) = \sigma(0,r+\delta) + \sigma(r+\delta,r),$$

and by 1)

$$\sigma(r+\delta,r) = \sigma(0, \frac{r-(r+\delta)}{1-(r+\delta)r}) = \sigma(0, \frac{-\delta}{1-r\delta-r^2}).$$

Choosing $\delta < 0$ with $|\delta|$ so small that $1-r\delta-r^2 > 0$, since $1 > 1-r(r+\delta) > 1-r^2 > 0$, we have, by 3)

$$\frac{1}{\delta}(\sigma(0,r+\delta) - \sigma(0,r)) = - \frac{1}{\delta}(\sigma(0,r) - \sigma(0,r+\delta)) =$$

$$= - \frac{1}{\delta}\sigma(r+\delta,r) = \frac{1}{1-r\delta-r^2} \frac{1-r\delta-r^2}{-\delta} \sigma(0, \frac{-\delta}{1-r\delta-r^2})$$

and

$$\lim_{\delta \nearrow 0} \frac{1-r\delta-r^2}{-\delta} \sigma(0, \frac{-\delta}{1-r\delta-r^2}) = 1$$

by 3). In conclusion

$$\lim_{\delta \to 0} \frac{1}{\delta}(\sigma(0,r+\delta) - \sigma(0,r)) = \frac{1}{1-r^2}$$

for all $0 \leqslant r < 1$. That proves that the function $r \to \sigma(0,r)$ is differentiable of class C^1 on $[0,1)$, and

$$\frac{d}{dr} \sigma(0,r) = \frac{1}{1-r^2} .$$

Thus

$$\sigma(0,r) = \frac{1}{2} \log \frac{1+r}{1-r} + \text{cost}.$$

Since by 3) $\sigma(0,0) = 0$, then

$$\sigma(0,r) = \frac{1}{2} \log \frac{1+r}{1-r} .$$

Thus, by 1),

$$\sigma(0,\zeta) = \sigma(0,|\zeta|) = \frac{1}{2}\log\frac{1+|\zeta|}{1-|\zeta|},$$

for all $z \in \Delta$, i.e., $\sigma(0,z) = \omega(0,z)$. Given ζ_1,ζ_2 in Δ, we have then, by 1),

$$\sigma(\zeta_1,\zeta_2) = \sigma\left(0, \frac{\zeta_2-\zeta_1}{1-\bar{\zeta}_1\zeta_2}\right) = \omega\left(0, \frac{\zeta_2-\zeta_1}{1-\bar{\zeta}_1\zeta_2}\right) = \omega(\zeta_1,\zeta_2).$$

Q.E.D.

IV. We will now construct the geodesic lines for the Poincaré hyperbolic metric. Let C be an orthogonal circle to $\partial\Delta$; i.e., C is a circle in $\mathbb{C}$ such that $C \cap \partial\Delta \neq \emptyset$ and on $C \cap \partial\Delta$ the two circles A and $\partial\Delta$ are orthogonal. If c is the center of C, then $|c| \geqslant 1$, and $\sqrt{|c|^2-1}$ is the radius of C. Thus

$$C = \{\zeta \in \mathbb{C}: |\zeta-c|^2 = |c|^2-1\} = \{\zeta \in \mathbb{C}: |\zeta|^2-2\,\mathrm{Re}(\zeta\bar{c})+1 = 0\}. \tag{A.11}$$

If $|c| = 1$, $C = \{c\}$, and viceversa. Hence $C \cap \Delta \neq \emptyset$, if, and only if, $|c| > 1$.

Lemma A.10. *If* $C \cap \Delta \neq \emptyset$, *the image of* $C \cap \Delta$ *by any* $g \in \mathrm{Aut}(\Delta)$ *is either the intersection of* Δ *with an orthogonal circle to* $\partial\Delta$, *or is a diameter of the circle* $\partial\Delta$. *The latter case arises, if, and only if,* $0 \in g(C \cap \Delta)$.

Proof. The lemma is trivial in case g is a rotation. Suppose then

$$g(\zeta) = \frac{\zeta+\zeta_0}{1+\bar{\zeta}_0\zeta}$$

with $\zeta_0 \in \Delta\setminus\{0\}$. For all $\zeta \in g(C \cap \Delta)$,

$$|g^{-1}(\zeta)-c|^2 = |c|^2-1$$

i.e.,

$$|\zeta-\zeta_0|^2-2\mathrm{Re}(\bar{c}(\zeta-\zeta_0)(1-\zeta_0\bar{\zeta}))+|1-\bar{\zeta}_0\zeta|^2 = 0,$$

or also

(A.12) $$(1+2\mathrm{Re}(\bar{c}\zeta_0)+|\zeta_0|^2)(|\zeta|^2+1)-2\mathrm{Re}((\overline{c+\bar{c}\zeta_0^2+2\bar{\zeta}_0})\zeta) = 0,$$

for all $\zeta \in g(C \cap \Delta)$.

If $1+2\mathrm{Re}(\bar{c}\zeta_0)+|\zeta_0|^2 \neq 0$, (A.12) can be written

$$|\zeta|^2-2\mathrm{Re}\left(\frac{\overline{c+c\bar{\zeta}_0^2+2\zeta_0}}{1+2\mathrm{Re}(\bar{c}\zeta_0)+|\zeta_0|^2}\cdot\zeta\right) + 1 = 0,$$

and this is the equation of a circle, which, by (A.11), is orthogonal to $\partial\Delta$. If $1+2\mathrm{Re}(\bar{c}\zeta_0)+|\zeta_0|^2$ vanishes - i.e., if $-\zeta_0 = g^{-1}(0) \in C$, hence $0 \in g(C)$ - then (A.12) reduces to

$$\mathrm{Re}((\bar{c} + \bar{c}\zeta_0^2 + 2\zeta_0)\zeta) = 0.$$

This equation - which cannot be identically satisfied (since the right hand side of (A.12) cannot vanish identically) - represents a straight line through 0.

Q.E.D.

Since the geodesic lines for the Poincaré hyperbolic metric at 0 are the radii of the circle $\partial\Delta$, Lemma A.10 yields

<u>Proposition A.11</u>. *The geodesic lines of the Poincaré hyperbolic metric on* Δ *are the diameters of the circle* $\partial\Delta$ *and the intersection of* Δ *with the orthogonal circles to* $\partial\Delta$.

As a consequence, every two distinct points of Δ are joined by a unique geodesic line.

We will now describe the balls

$$B_\omega(\zeta_0,d) = \{\zeta \in \Delta : \omega(\zeta_0,\zeta) < d\}$$

with center $\zeta_0 \in \Delta$ and radius d, for the Poincaré distance.

Since the function $t \mapsto \log\frac{1+t}{1-t}$ is strictly increasing on the interval $[0,d)$, then

$$B_\omega(\zeta_0,d) = \{\zeta \in \Delta : \left|\frac{\zeta-\zeta_0}{1-\bar{\zeta}_0\zeta}\right| < r\}$$

where r, with $0<r<1$, is the unique solution of the equation

$$\frac{1}{2}\log\frac{1+r}{1-r} = d,$$

and is given by

$$r = \frac{e^{2d}-1}{e^{2d}+1}.$$

By simple manipulations, the equation

$$\left|\frac{\zeta-\zeta_0}{1-\bar{\zeta}_0\zeta}\right| = r \tag{A.13}$$

becomes

$$\left|\zeta - \frac{1-r^2}{1-r^2|\zeta_0|^2}\zeta_0\right| = \frac{r(1-|\zeta_0|^2)}{1-r^2|\zeta_0|^2},$$

and thus represents a circle with center $\zeta_1 = \frac{1-r^2}{1-r^2|\zeta_0|^2}\zeta_0$ and radius $\rho = \frac{r(1-|\zeta_0|^2)}{1-r^2|\zeta_0|^2}$. Being

$$|\zeta_1|+\rho-1 = \frac{(1-r)(1-|\zeta_0|)(r|\zeta_0|-1)}{1-r^2|\zeta_0|^2} < 0$$

the circle (A.13) is contained in Δ. In conclusion we have proved

Proposition A.12. *The disc* $B_\omega(\zeta_0,d)$ *is the disc - for the Euclidean distance in* $\mathbb{C}$ *- with center* $\zeta_1 = \frac{2}{1-r^2|\zeta_0|^2}\zeta_0$ *and radius* $\rho = r\frac{1-|\zeta_0|^2}{1-r^2|\zeta_0|^2}$. *Furthermore* $\overline{B_\omega(\zeta_0,d)} \subset \Delta$.

As a consequence, $\overline{B_\omega(\zeta_0,d)}$ is compact, hence sequentially compact, and that implies that the Poincaré distance is complete.

V. Let $\Pi^+ = \{\zeta \in \mathbb{C}: \operatorname{Im}\zeta > 0\}$. The function

$$C: \zeta \mapsto \frac{\zeta - i}{\zeta + i}$$

is holomorphic on Π^+. Being

$$(A.14) \qquad 1 - |C(\zeta)|^2 = \frac{4 \cdot \operatorname{Im} \zeta}{|\zeta + i|^2} > 0 \quad \text{for} \quad \zeta \in \Pi^+,$$

then $C(\Pi^+) \subset \Delta$. Since

$$\zeta = i \, \frac{1 + C(\zeta)}{1 - C(\zeta)} \, ,$$

and

$$\operatorname{Im} \zeta = \frac{1 - |C(\zeta)|^2}{|1 - C(\zeta)|^2} \, ,$$

then C is a conformal map of Π^+ onto Δ. Being $C(i) = 0$,

$$(A.15) \qquad \frac{dC}{d\zeta} = \frac{2i}{(\zeta + 1)^2} \qquad (\zeta \in \Delta)$$

and in particular

$$\left(\frac{dC}{d\zeta} \right)_{\zeta = i} = \frac{-i}{2} \, ,$$

then, by the Riemann conformal mapping theorem, C is the unique conformal map of Π^+ onto Δ, mapping i onto 0, whose derivative at i is purely imaginary, with a negative imaginary coefficient. C is called the *Cayley transform*.

By (A.14) and (A.15) the image of the Poincaré hyperbolic metric on Δ is the Riemannian metric on Π^+ (called the Poincaré hyperbolic metric on Π^+):

$$ds^2 = \frac{1}{4(\operatorname{Im} \zeta)^2} |d\zeta|^2 \qquad (\zeta \in \Pi^+).$$

Similarly, being

$$\frac{C(\zeta_1) - C(\zeta_2)}{1 - \overline{C(\zeta_1)} C(\zeta_2)} = \frac{\overline{(\zeta_1 + i)}}{\zeta_1 + i} \; \frac{\zeta_1 - \zeta_2}{\bar{\zeta}_1 - \zeta_2} \qquad (\zeta_1, \zeta_2 \in \Pi^+),$$

the *Poincaré distance* on Π^+ is given by

$$\omega(C(\zeta_1),C(\zeta_2)) = \frac{1}{2}\log \frac{1 + \left|\dfrac{\zeta_1-\zeta_2}{\bar{\zeta}_1-\zeta_2}\right|}{1 - \left|\dfrac{\zeta_1-\zeta_2}{\bar{\zeta}_1-\zeta_2}\right|} \qquad (\zeta_1,\zeta_2 \in \Pi^+).$$

All the results obtained in I-IV yield - via the Cayley transform and trivial computations - similar results for the Poincaré hyperbolic metric and the Poincaré distance. For example:

The geodesic lines for the Poincaré hyperbolic metric on Π^+ *are all the intersections of* Π^+ *with vertical straight lines and with circles centered on the real axis.*

The balls for the Poincaré distance on Π^+ *are all the discs (for the euclidean distance) whose closures are contained in* Π^+.

Notes.

For the geometry of the Poincaré metric cf., e.g., [Bianchi, 1; Vol.I, Parte II, cap. XIV, pp. 607-654]. Concerning the arguments leading to Proposition A.7, cf., e.g., [Bianchi, 1; Vol.I, Parte I, § 49, pp. 123-129].

Proposition A.9 can be found in [Hille, 1; pp. 238-239]. For a different characterization of the Poincaré metric, cf. [Eisenman, 1; Proposition 2.18, pp. 13-16].

APPENDIX B

BAIRE SPACES

Let X be a topological space. To avoid trivialities, we suppose $X \neq \emptyset$. A subset C of X is called: a *rare* (or a *nowhere dense*) subset if $\mathring{\overline{C}} = \emptyset$; a *meager* subset (or a set *of the first category*) if it is a countable union of rare subsets; a *residual* set if its complement $X \setminus C$ is meager; a set *of the second category* if it is not meager.

The proof the following lemma is left as an exercise:

<u>Lemma B.1</u>. *A subset of* X *is meager if, and only if, it is contained in a countable union of closed rare sets.*

<u>Lemma B.2</u>. *Let* Y *be a subspace of* X*, and let* $C \subset Y$.

i. If C *is rare in* Y*, then* C *is rare in* X.

ii. If Y *is open in* X*, and* C *is rare in* X*, then* C *is rare in* Y.

<u>Proof</u>. i) Let $\overline{C}$ be the closure of C in X, and let A be a non-empty open set in X such that $A \subset \overline{C}$. Then $A \cap C \neq \emptyset$, i.e., $(A \cap Y) \cap C = A \cap C \neq \emptyset$, and therefore $A \cap Y$ is a non-empty open subset of Y. Thus the closure $\overline{C}_Y$ of C in Y, which coincides with $\overline{C} \cap Y$, contains also the non-empty open set $A \cap Y$.

ii) Suppose that there is a non-empty open subset B of

Y such that $B \subset \overline{C}_Y = \overline{C} \cap Y$.

B is open in X. Thus $B \subset \overline{C}$, and this contradiction completes the proof of the lemma.

Q.E.D.

<u>Lemma B.3</u>. *The following four conditions are equivalent:*

α) *Every countable intersection of everywhere dense open sets is everywhere dense.*

β) *Every countable union of rare closed sets has empty interior.*

γ) *Every non-empty open set is not meager.*

δ) *Every residual set is everywhere dense.*

<u>Proof</u>. α) ⇒ β). Let (F_n) be a sequence of closed subsets of X, and let $S = \cup F_n$. If $\overset{\circ}{F}_n = \emptyset$, i.e., if the open sets $X \setminus F_n$ are everywhere dense in X, then $X \setminus S = X \setminus (\cup F_n) = \cap (X \setminus F_n)$ is everywhere dense in X, i.e., $\overset{\circ}{S} = \emptyset$.

β) ⇒ γ). Follows from Lemma B.1.

γ) ⇒ δ). Let C be residual, i.e., let $X \setminus C$ be meager. If $\overline{C} \neq X$, then $X \setminus \overline{C}$ is open, and therefore is not meager. But this is a contradiction, for $X \setminus C \subset X \setminus C$, and every subset of a meager set is meager.

δ) ⇒ α). Let (A_n) be a sequence of open sets and let $C = \cap A_n$. If $\overline{A}_n = X$, i.e., if the closed sets $X \setminus A_n$ are rare, then $X \setminus C = X \setminus (\cap A_n) = (X \setminus A_n)$ is meager, i.e., C is residual.

Q.E.D.

<u>Definition</u>. A topological space satisfying any one of conditions α)-δ) is called a *Baire space*.

The following famous theorem of Baire provides important examples of Baire spaces.

Theorem B.4. *If the topological space* X *is either*

i) a complete metric space,

ii) a locally compact (Hausdorff) space,

then X *is a Baire space.*

Proof. Let (A_n) be a sequence of open dense subsets of X, and let B be any non-empty open set in X.

We shall show that, letting

$$C = \cap A_n ,$$

then $B \cap C \neq \emptyset$.

We consider first the case i): let d be a distance on X, let $B(x,r) = \{y \in X: d(x,y) < r\}$ and let $\overline{B(x,r)}$ be the closure of $B(x,r)$.

Since A_1 is dense, the open set $A_1 \cap B$ is non-empty. Hence there is some $x_1 \in X$ and some $r_1 > 0$, such that $\overline{B(x_1,r_1)} \subset A_1 \cap B$. Clearly, we can choose $0 < r_1 < 1$. Since A_2 is dense, the open set $B(x_1,r_1) \cap A_2$ is non-empty. Hence, there is some $x_2 \in X$ and some r_2 such that $\overline{B(x_2,r_2)} \subset A_2 \cap \cap B(x_1,r_1)$, and $0 < r_2 < 1/2$. Iterating this procedure, suppose that $x_1, \dots, x_{n-1}$ and $r_1, \dots, r_{n-1}$ are chosen. Since A_n is dense, the open set $B(x_{n-1},r_{n-1}) \cap A_n$ is non-empty. Hence there is some $x_n \in X$ and some r_n such that

$$\text{(B.1)} \qquad \overline{B(x_n,r_n)} \subset A_n \cap B(x_{n-1},r_{n-1}),$$

and we can choose $0 < r_n < \frac{1}{n}$.

The sequence (x_n) is a Cauchy sequence in X, for - given any index n - whenever $p,q > n$, then $x_p , x_q \in B(x_n ,r_n)$, and therefore

$$d(x_p ,x_q) \leqslant d(x_p ,x_n) + d(x_n ,x_q) < \frac{1}{n} + \frac{1}{n} = \frac{2}{n} .$$

Since X is complete, the sequence (x_n) converges to some point $x \in X$. Since $x_p \in B(x_n, r_n)$ for all $p \geqslant n$, then $x \in \overline{B(x_n, r_n)}$ for all indices n. Thus, by (B.1),

$$x \in \cap \overline{B(x_n, r_n)} \in \cap (A_n \cap B),$$

showing that $B \cap C \neq \emptyset$, and thereby proving the theorem under condition i).

If hypothesis ii) holds, we choose a non-empty open set B_1 such that $\overline{B}_1$ is compact and $\overline{B}_1 \subset A_1 \cap B$. Similarly, we choose a non-empty open set B_2 such that $\overline{B}_2 \subset A_2 \cap B_1, \dots$, a non-empty open set B_n such that $\overline{B}_n \subset A_n \cap B_{n-1}$.

Let $K = \cap \overline{B}_n$. Then $K \neq \emptyset$ for $\overline{B}_n \subset B_{n-1}$ and $\overline{B}_1$ is compact. Since $K \subset B \cap A_n$ for every n, that shows that $C \cap B \neq \emptyset$.

Q.E.D.

Proposition B.5. *Any open non-empty subset of a Baire space is a Baire space for the relative topology.*

Proof. Let Y be a non-empty open subset of the Baire space X, and let (F_n) be a sequence of closed subsets of Y. Let U be a non-empty open subset of Y such that $U \subset \cup F_n$.

U being open in the Baire space X, and being $U \subset \cup \overline{F}_n$, then $\overset{\circ}{\overline{F}}_{n_0} \neq \emptyset$ for some index n_0, i.e., there is a non-empty open set A in X such that $A \subset \overline{F}_{n_0}$ and therefore $A \cap F_{n_0} \neq \emptyset$.

Thus F_{n_0} contains the non-empty open subset $A \cap Y$ of Y.

Q.E.D.

Exercise. Prove that, if the topological space X is a union of open sets, each one of which is a Baire space for the relative topology, then X is a Baire space.

This fact, together with Proposition B.5, implies that the property of being a Baire space is a local property.

Example. If X is either a complete metric space or a locally compact (Hausdorff) space, and Y is a closed subset of X, then Y is a space of the same type as X, for the relative topology. So, each closed subset of a complete metric space, or of a locally compact (Hausdorff) space is itself (by Theorem B.4) a Baire space. However, not every closed subset of any Baire space is a Baire space for the relative topology.

Let $X = \mathbb{R}$, and consider on X the following topology: a subset A of X is open if, and only if, it is the union of an open set for the usual topology of $\mathbb{R}$, and of a subset of $\mathbb{R} \setminus \mathbb{Q}$.

It is not difficult to prove that:

1) X is a Baire space;

2) $\mathbb{Q}$ is a closed subset of X, which is not a Baire space for the relative topology (which is the same as the relative topology of $\mathbb{Q}$ in $\mathbb{R}$).

Let Y be a metric space with a distance d. For any map $f: X \to Y$ we define the *oscillation* $o(f:x)$ of f at $x \in X$ by $o(f:x) = \inf\{\text{diam } f(U): U \text{ neighborhood of } x\}$.

Obviously, the map f is continuous if, and only if, $o(f:x) = 0$.

Theorem B.6. (W.F. Osgood) *Let* X *be a Baire space and let* (f_n) *be a sequence of continuous maps of* X *into a metric space* Y. *If the sequence* $(f_n(x))$ *converges at each point*

$x \in X$, *the map* $f: X \to Y$ *defined by* $f(x) = \lim_{n\to\infty} f_n(x)$ *at each point* $x \in X$ *is continuous on a residual set.*

Proof. 1) Let n be a positive integer, and, for $p=1,2,\dots,$ let

$$H_p = \{x \in X: d(f_p(x), f_{p+q}(x)) \leqslant \frac{1}{4n} \quad \text{for } q=0,1,2,\dots\}$$

The function f_{p+q} being continuous, then $x \to d(f_p(x), f_{p+q}(x))$ is a continuous function on X, and therefore H_p is closed. Since $\lim f_p(x) = f(x)$ exists at each point of X, then $\cup H_p = X$. Thus, by Lemma B.3, there is some p for which $\overset{\circ}{H}_p \neq \emptyset$.

2) By Proposition B.5, the conclusion of 1) holds for any open subspace of X. Thus for every positive integer n there exists a dense open set A_n in X with the following property: for every $x \in A_n$ there is some index p and a neighborhood U_p of x in A_p such that

$$\text{(B.2)} \qquad d(f_p(y), f_{p+q}(y)) \leqslant \frac{1}{4n} \quad \text{for all } q=0,1,2,\dots \text{ and all } y \in U_p.$$

Since X is a Baire space, the residual set $A = \cap A_n$ is dense in X. We shall prove the theorem by showing that, for every $x \in A$, $o(f:x) = 0$.

Let $\varepsilon > 0$, and let n be a positive integer with $\frac{1}{n} < \varepsilon$. Let p_0 be such that (B.2) holds with $p=p_0$, and let U be a neighborhood of x for which $U \subset U_{p_0} \subset A$, and $\operatorname{diam} f_{p_0}(U) < \frac{1}{4n}$.

Then, for $x', x'' \in U$,

$$d(f(x'), f(x'')) = \lim_{p,q\to\infty} d(f_p(x'), f_q(x'')) \leqslant$$

$$\leqslant \lim_{p,q\to\infty} (d(f_p(x'), f_{p_0}(x')) + d(f_{p_0}(x'), f_{p_0}(x)) + d(f_{p_0}(x), f_{p_0}(x'')) +$$

$$+ d(f_{p_0}(x''), f_q(x'')) \leqslant \frac{1}{4n} + \frac{1}{4n} + \frac{1}{4n} + \frac{1}{4n} = \frac{1}{n} < \varepsilon .$$

Q.E.D.

By Lemma B.3, Theorem B.6 implies that the map f is continuous on a dense subset of X. However, the map f may be discontinuous on a dense subset.

Example. (cf. [Munroe, 1, pp.70-71]). Let (x_k) be a dense sequence of real numbers, and let $g_{kn}: \mathbb{R} \to \mathbb{R}$ be defined by

$$g_{kn}(t) = \begin{cases} 0 & \text{for } t < x_k , \\ n(t-x_k) & \text{for } x_k \leqslant t \leqslant x_k + \frac{1}{n} , \\ 1 & \text{for } t > x_k + \frac{1}{n} \end{cases}$$

Let

$$f_n(t) = \sum_{k=1}^{n} \frac{1}{2^k} g_{kn}(t).$$

The sequence (f_n) is monotone and bounded, hence convergent at each point of $\mathbb{R}$. However the limit function is discontinuous at each point x_k .

Proposition B.5 yields the following generalization of Theorem B.6.

Theorem B.7. *Let* (f_n) *be a sequence of maps of a Baire space* X *into a metric space* Y, *such that* $(f_n(x))$ *converges to a point* $f(x)$ *at every* $x \in X$. *If each* f_n *is continuous on a residual set, then also the map* $f: X \to Y$ *is continuous on a residual set.*

We shall now discuss real valued functions on a Baire space X. Recall that a function $f: X \to \mathbb{R}$ is lower semi-continuous (resp. upper semi-continuous) if, for any $c \in \mathbb{R}$, the set $\{x \in X: f(x) > c\}$ (resp.: the set $\{x \in X: f(x) < c\}$)

is open.

<u>Lemma B.8</u>. *Let* X *be a Baire space and let* $f: X \to \mathbb{R}$ *be a lower semi-continuous function on* X. *Then* f *is bounded from above on a non-empty open set in* X.

<u>Proof</u>. Since f is lower semi-continuous, the sets $X_n = \{x \in X: f(x) \leq n\}$ are closed for $n=1,2, \dots$. Being $f(x) < +\infty$ for all $x \in X$, then $X = \cup X_n$. Thus $\overset{\circ}{X}_n \neq \emptyset$ for some n.

Q.E.D.

If X is a complete metric space, and the lower semi-continuous function in Lemma B.8 is bounded from below, much more can be said.

We begin by proving the following

<u>Lemma B.9</u>. *Let* X *be a metric space with a distance* d, *and let* $f: X \to \mathbb{R}^+ \cup \{+\infty\}$ $(= \overline{\mathbb{R}}^+)$ *be a lower semi-continuous function. Then, there exists a sequence* (f_n) *of real valued continuous functions on* X *such that for each* $x \in X$

$$f_n(x) \nearrow f(x) \quad \textit{as} \quad n \to \infty .$$

<u>Proof</u>. 1) If $f(x) = +\infty$ for all $x \in X$, the proof is trivial. Therefore, we can assume $f \not\equiv +\infty$. We define f_n by setting

$$f_n(x) = \inf\{f(y) + nd(x,y): y \in X\}.$$

Clearly, $f_n(x) < +\infty$ for all $x \in X$, and

$$f_n(x) \leq f(x) + nd(x,x) = f(x). \tag{B.3}$$

Keeping x and y fixed, the numerical sequence $(f(y) + nd(x,y))$ is not decreasing as $n \to \infty$. Thus, for every

$x \in X$ the sequence $(f_n(x))$ is not decreasing, as $n \to \infty$. Since $d(x,y) \leqslant d(x,z) + d(z,y)$ $(x,y,z \in X)$, then $f_n(x) \leqslant$ $\leqslant f(y) + nd(x,y) \leqslant f(y) + nd(x,z) + nd(z,y)$. Therefore

$$f_n(x) \leqslant f_n(z) + nd(x,z),$$

and also

$$f_n(z) \leqslant f_n(x) + nd(z,x),$$

so that

$$|f_n(x) - f_n(z)| \leqslant nd(x,z).$$

That shows that f_n is a (uniformly) continuous function.

2) We prove now that $f_n(x) \nearrow f(x)$ as $n \to \infty$.

Let $t < f(x)$. Since f is lower semi-continuous, the set $X(t) = \{y \in X: f(y) > t\}$ is open in X. Being $x \in X(t)$, there is some $r > 0$ such that the open ball $B(x,r)$ with center x and radius r is contained in $X(t)$. Choose an integer n_0 in such a way that $n_0 r > t$. If $y \in X(t)$, then

$$f(y) + nd(x,y) \geqslant f(y) > t \qquad \text{for all } n \geqslant 0,$$

while, if $y \notin X(t)$,

$$f(y) + nd(x,y) \geqslant nd(x,y) \geqslant nr \geqslant n_0 r > t \qquad \text{for all } n \geqslant n_0.$$

Hence, taking into account (B.3), we have

$$t \leqslant f_n(x) \leqslant f(x) \qquad \text{for all } n \geqslant n_0.$$

That shows that $f_n(x) \nearrow f(x)$.

Q.E.D.

The above Lemma and Theorem B.6 yield

Theorem B.10. *If* X *is a complete metric space, every lower semi-continuous real valued function, bounded from below on* X, *is continuous on a residual set.*

Passing from lower semi-continuous to upper semi-continuous functions we have

Lemma B.11. *Every upper semi-continuous function* $f: X \to \mathbb{R}$ *on a Baire space* X *is bounded from below on a non-empty open set in* X.

Theorem B.12. *If* X *is a complete metric space, every upper semi-continuous function bounded from above on* X *is continuous on a residual set.*

Finally, let $(f_i)_{i \in I}$ be a family of lower semi-continuous functions on X, whose upper envelope $x \mapsto f(x) = \sup\{f_i(x) \mid i \in I\}$ is finite at each $x \in X$.

The function f satisfies all the hypotheses of Lemma B.8, which, together with Proposition B.5, yields

Proposition B.13. *Let* (f_i) *be a family of lower semi-continuous functions on a Baire space* X, *such that* $\sup\{f_i(x) : i \in I\} < +\infty$ *for each* $x \in X$. *Then every non-empty open set in* X *contains a non-empty open set on which the family* (f_i) *is uniformly bounded from above.*

Notes.

Further information on Baire spaces can be found in [Bourbaki, 1], [Holmes, 1] [Kuratowski, 1], [Munroe, 1].

REFERENCES

[Ahlfors, 1] Ahlfors, L.V., Complex Analysis (Mc-Graw Hill, New York, 1966).

[Ahlfors, 2] Ahlfors, L.V., Conformal Invariants. Topics in Geometric Function Theory (Mc-Graw Hill, New York, 1973).

[Akeman-Russo, 1] Akeman, C.A. and Russo, B., Geometry of the unit sphere of a $C^\star$-algebra and its dual, Pacific J. Math. 32 (1970) 575-585.

[Andreotti-Vesentini, 1] Andreotti, A. and Vesentini, E., On deformations of discontinuous groups, Acta Math. 112 (1964) 249-298.

[Banach, 1] Banach, S., Théorie des opérations linéaires, Monografje Matematyczne, Warsaw, Tom I (1932).

[Barth, 1] Barth, T.J., The Kobayashi distance induces the standard topology, Proc. Amer. Math. Soc. 35 (1972) 439-441.

[Behnke-Thullen, 1] Behnke, H. and Thullen, P., Theorie der Funktionen mehrerer komplexer Veränderlichen, 2nd edition (Springer-Verlag, Berlin-Heidelberg-New York, 1970).

[Bergman, 1] Bergman, S., The kernel function and the conformal mapping, Mathematical Surveys, N.5, Amer. Math. Soc., Providence, R.I. (1950).

[Bianchi, 1] Bianchi, L., Lezioni di Geometria Differenziale, 3 ed. (Zanichelli, Bologna, 1927).

[Bochner-Martin, 1] Bochner, S. and Martin, W.T., Several Complex Variables (Princeton University Press, Princeton, 1948).

[Bohnenblust-Karlin, 1] Bohnenblust, H.F. and Karlin, S., Geometrical properties of the unit sphere of Banach algebras, Ann. of Math. 62 (1955) 217-229.

[Bourbaki, 1] Bourbaki, N., Eléments de Mathématique, Topologie Générale, Chapitre IX (Hermann, Paris, 1948).

[Bourbaki, 2] Bourbaki, N., Eléments de Mathématique, Théories Spectrales, Chapitres 1 et 2 (Hermann, Paris, 1967).

[Braun-Kaup-Upmeier, 1] Braun, R., Kaup, W. and Upmeier, H., On the automorphisms of circular and Reinhardt domains in complex Banach spaces, Manuscripta Math. 25 (1978) 97-133.

[Brown-Douglas, 1] Brown, A. and Douglas, R.G., On maximum theorems for analytic operator functions, Acta Sci. Math. (Szeged) 26 (1966) 325-327.

[Carathéodory, 1] Carathéodory, C., Über das Schwarzsche Lemma bei analytischen Funktionen von zwei komplexen Veränderlichen, Math. Ann. 97 (1926) 76-98; Gesammelte Mathematische Schriften, vol.IV, pp.132-

159 (C.H. Beck'sche Verlagsbuchhandlung, München, 1956).

[Carathéodory, 2] Carathéodory, C., Über eine spezielle Metrik, die in der Theorie des Analytischen Funktionen auftritt, Atti della Pontificia Accademia delle Scienze Nuovi Lincei 80 (1927) 135-141; Gesammelte Mathematische Schriften, vol.IV, pp.160-166.

[Carathéodory, 3] Carathéodory, C., Über die Geometrie der analytische Abbildungen, die durch analytische Funktionen von zwei Veränderlichen vermittelt werden, Abhandl. Math. Sern. Univ. Hamburg 6 (1928) 97-145; Gesammelte Mathematische Schriften, vol.IV, 167-227.

[Cartan, 1] Cartan, H., Les fonctions de deux variables complexes et le problème de la représentation analytique, J. Math. Pures Appl. (9) 10 (1931) 1-114.

[Cartan, 2] Cartan, H., Théorie Elémentaire des Fonctions Analytiques d'une ou plusieurs Variables Complexes (Hermann, Paris, 1961).

[Dieudonné, 1] Dieudonné, J., Eléments d'Analyse. 1. Fondements de l'Analyse Moderne (Gauthier-Villars, Paris, 1972).

[Douady, 1] Douady, A., Le problème des modules pour les sous-espaces analytiques compacts d'un espace analytique donné, Ann. Inst. Fourier (Grenoble) 16

(1966) 1-95.

[Dunford-Schwartz, 1] Dunford, N. and Schwartz, J.T., Linear Operators, Part I (Interscience Publishers Inc., New York, 1958).

[Earle-Hamilton, 1] Earle, C.J. and Hamilton, R.S., A fixed point theorem for holomorphic mappings, Proc. Symposia Pure Math., Amer. Math. Soc., Providence, R.I., vol. 16 (1979) 61-65.

[Eisenman, 1] Eisenman, D.A., Intrinsic measures on complex manifolds and holomorphic mappings, Mem. Amer. Math. Soc. 96 (1970) 1-80.

[Goluzin, 1] Goluzin, G.M., Geometric theory of functions of a complex variable, Transl. of Math. Monographs, vol.26, Amer. Math. Soc., 1969.

[Greenfield-Wallach, 1] Greenfield, S. and Wallach, N., Automorphism groups of bounded domains in Banach spaces, Trans. Amer. Math. Soc. 166 (1972) 45-57.

[Hahn, 1] Hahn, K.T., On completeness of the Bergman metric and its subordinate metric, Proc. Nat. Acad. Sci. U.S.A. 73 (1976) 4294.

[Halmos, 1] Halmos, P.R., A Hilbert Space Problem Book (Van Nostrand, Princeton, 1967).

[Harris, 1] Harris, L.A., Schwarz's lemma in normed linear spaces, Proc. Nat. Acad. Sci.

U.S.A. 62 (1969) 1014-1017.

[Harris, 2] Harris, L.A., Bounded symmetric homogeneous domains in infinite dimensional spaces, Lecture Notes in Mathematics, # 364, 13-40 (Springer-Verlag, Berlin-Heidelberg-New York, 1973).

[Harris, 3] Harris, L.A., Operator extreme points and the numerical range, Indiana Univ. Math. J. 23 (1974) 937-947.

[Harris, 4] Harris, L.A., Operator Siegel domains, Proc. Roy. Soc. Edinburgh Sect. A 79 (1977) 137-156.

[Hayden-Suffridge, 1] Hayden, T.L. and Suffridge, T.J., Biholomorphic maps in Hilbert space have a fixed point, Pacific J. Math. 38 (1971) 419-422.

[Hayden-Suffridge, 2] Hayden, T.L. and Suffridge, T.J., Fixed points of holomorphic maps in Banach spaces, Proc. Amer. Math. Soc. 60 (1976) 95-105.

[Helgason, 1] Helgason, S., Differential Geometry and Symmetric Spaces (Academic Press, New York, 1962).

[Hervé, 1] Hervé, M., Several Complex Variables. Local Theory (Oxford University Press, 1963).

[Hervé, 2] Hervé, M., Quelques propriétés des applications analytiques d'une boule à m dimensions dans elle même, J.

Math. Pures Appl. (9) 42 (1963) 117-147.

[Hervé, 3] Hervé, M., Analytic and plurisubharmonic functions in finite and infinite dimensional spaces, Lecture Notes in Mathematics, # 198 (Springer-Verlag, Berlin-Heidelberg-New York, 1971).

[Hervé, 4] Hervé, M., Some properties of the images of analytic maps, in: Matos, M.C. (ed.), Infinite Dimensional Holomorphy and Applications (North-Holland, Amsterdam, 1977).

[Hille, 1] Hille, E., Analytic Function Theory, vol.II (Ginn and Co., Boston, 1962).

[Hille-Phillips, 1] Hille, E. and Phillips, R.S., Functional analysis and semi-groups, Amer. Math. Soc. Colloquium Publ. vol.36, Amer. Math. Soc., Providence, R.I., 1957.

[Hirzebruch, 1] Hirzebruch, U., Halbräume und ihre holomorphen Automorphismen, Math. Annalen 153 (1964) 395-417.

[Holmes, 1] Homes, R.B., Geometric Functional Analysis and its Application (Springer-Verlag, Berlin-Heidelberg-New York, 1975).

[Hörmander, 1] Hörmander, L., An Introduction to Complex Analysis in Several Variables (Van Nostrand, Princeton, 1966).

[Hua, 1] Hua, L.K., Harmonic analysis of functions of several complex variables in the classical domains, Amer.Math.Soc., Transl. of Math. Monographs, vol.6, 1963.

[Kadison, 1] Kadison, R.V., Isometries of operator algebras, Ann. of Math. 54 (1951) 325-338.

[Kakutani, 1] Kakutani, S., Topological properties of the unit sphere of a Hilbert space, Proc. Imp. Acad.,Tokyo 19 (1943) 269-271.

[Kaup, 1] Kaup, W., Bounded symmetric domains in finite and infinite dimensions. A review, in: Several Complex Variables, Cortona, Italy, 1976-77 (Scuola Normale Superiore, Pisa, 1978).

[Kaup-Upmeier, 1] Kaup, W. and Upmeier, H., Banach spaces with biholomorphically equivalent unit balls are isomorphic, Proc. Amer. Math. Soc. 58 (1976) 129-133.

[Klingen, 1] Klingen, H., Über die analytischen Abbildungen verallgemeinerter Einheitskreise auf sich, Math. Ann. 132 (1956) 134-144.

[Klingen, 2] Klingen, H., Analytic automorphisms of bounded symmetric complex domains, Pacific J. Math. 10 (1960) 1327-1332.

[Kobayashi, 1] Kobayashi, S., Invariant distances on

complex manifolds and holomorphic mappings, J. Math. Soc.,Japan 19 (1967) 460-480.

[Kobayashi, 2] Kobayashi, S., Distance, holomorphic mappings and the Schwarz lemma, J. Math. Soc.,Japan 19 (1967) 481-485.

[Kobayashi, 3] Kobayashi, S., Hyperbolic Manifolds and Holomorphic Mappings (Marcel Dekker, New York, 1970).

[Kobayashi, 4] Kobayashi, S., Some remarks and questions concerning the intrinsic distance, Tohoku Math. J. 25 (1973) 481-486.

[Kobayashi, 5] Kobayashi, S., Intrinsic distances, measures and geometric function theory, Bull. Amer. Math. Soc. 82 (1976) 357-416.

[Kuratowski, 1] Kuratowski, C., Topologie I, Espaces Metrisables, Espaces Complets (Warszawa-Lwow, 1933).

[Lumer, 1] Lumer, G., Semi-inner-product spaces, Trans. Math. Soc. 100 (1961) 29-43.

[Mazur-Orlicz, 1] Mazur, S. and Orlicz, W., Grundlegende Eigenschaften der polynomischen Operationen, Studia Math. 5 (1934) 50-68, 179-189.

[McGuigan, 1] McGuigan, R., Strongly extreme points in Banach spaces, Manuscripta Math. 5 (1971) 113-122.

[Munroe, 1] Munroe, M.E., Introduction to Measure and Integration, I ed. (Addison-Wesley, Reading, Mass., 1953).

[Nachbin, 1] Nachbin, L., Topology on Spaces of Holomorphic Mappings (Springer-Verlag, Berlin-Heidelberg-New York, 1969).

[Nachbin, 2] Nachbin, L., A glimpse at infinite dimensional holomorphy, in: Lecture Notes in Mathematics, # 364 (Springer-Verlag, Berlin-Heidelberg-New York, 1974).

[Nachbin, 3] Nachbin, L., Warum unendlich dimensionale Holomorphie?, Jahrbuch Überblicke Mathematik, 1979 (Bibliographisches Institut, Mannheim-Zürich).

[Narasimhan, 1] Narasimhan, R., Several Complex Variables, Chicago Lectures in Mathematics (The University of Chicago Press, Chicago, 1971).

[Noverraz, 1] Noverraz, Ph., Fonctions plurisousharmôniques et analytiques dans les espaces vectoriels topologiques, Ann. Inst. Fourier (Grenoble) 19 (1969) 419-493.

[Noverraz, 2] Noverraz, Ph., Pseudo-convexité, Convexité Polynomiale et Domaines d'Holomorphie en Dimension Infinie (North Holland, Amsterdam, 1973).

[Phillips, 1] Phillips, R.S., On symplectic mappings of contraction operators, Studia Math. 31 (1968) 15-27.

[Rado, 1] Rado, T., Subharmonic Functions (Springer, Berlin, 1937).

[Ramis, 1] Ramis, J.-P., Sous-ensembles Analytiques d'une Variété Banachique Complexe (Springer-Verlag, Berlin-Heidelberg-New York, 1970).

[Reiffen, 1] Reiffen, H.J., Die differentialgeometrischen Eigenschaften der invarianten Distanzfunktion von Carathéodory, Schr. Math. Inst. Univ. Münster, N.26 (1963).

[Reiffen, 2] Reiffen, H.J., Die Carathéodorische Distanz und ihre zugehörige Differentialmetrik, Math. Ann. 161 (1965) 315-324.

[Renaud, 1] Renaud, A., Quelques propriétés des applications analytiques d'une boule de dimension infinie dans une autre, Bull. Sci. Math. (2) 97 (1973) 129-159.

[Rinow, 1] Rinow, W., Die innere Geometrie der metrischen Räume (Springer-Verlag, Berlin-Heidelberg-New York, 1961).

[Royden, 1] Royden, H.L., Remarks on the Kobayashi metric, Several complex variables, II. Lecture Notes in Mathematics, vol. 185 (Springer-Verlag, Berlin-Heidelberg-New York, 1971).

[Royden, 2] Royden, H.L., The extension of regular holomorphic maps, Proc. Amer. Math.

Soc. 43 (1974) 306-310.

[Rudin, 1] Rudin, W., Real and Complex Analysis (McGraw Hill, New York, 1966).

[Rudin, 2] Rudin, W., Functional Analysis (McGraw Hill, New York, 1973).

[Sakai, 1] Sakai, S., $C^{\star}$-algebras and $W^{\star}$-algebras (Springer-Verlag, Berlin-Heidelberg-New York, 1971).

[Siegel, 1] Siegel, C.L., Symplectic geometry, Amer. J. Math. 65 (1943) 1-86 and (Academic Press, New York and London, 1964); Gesammelte Abhandlungen, vol.II (Springer-Verlag, Berlin-Heidelberg-New York, 1966).

[Siegel, 2] Siegel, C.L., Analytic Functions of Several Complex Variables (Mimeographed Notes, Princeton, 1948).

[Siegel, 3] Siegel, C.L., Topics in Complex Function Theory, vol.III (John Wiley, New York, 1973).

[Stacho, 1] Stacho, J.L., On the existence of fixed points for holomorphic automorphisms, to appear.

[Thorp-Whitley, 1] Thorp, E. and Whitley, R., The strong maximum modulus theorem for analytic functions into a Banach space, Proc. Amer. Math. Soc. 18 (1967) 640-646.

[Upmeier, 1] Upmeier, H., Über die Automorphismengruppen beschränkter Gebiete in Banachräumen, Dissertation, Universität zu Tübingen (1975).

[Vesentini, 1] Vesentini, E., Maximum theorem for vector-valued holomorphic functions, Rend. Sem. Mat. Fis.,Milano 40 (1970) 24-55.

[Vesentini, 2] Vesentini, E., Automorphisms of the unit ball, in: Several Complex Variables, Cortona, Italy, 1976-1977 (Scuola Normale Superiore, Pisa, 1978).

[Vesentini, 3] Vesentini, E., Variations on a theme of Carathéodory, Ann. Scuola Norm. Sup. Pisa (4) 7 (1979).

[Vesentini, 4] Vesentini, E., Invariant distances and invariant differential metrics in locally convex spaces, Proc. Stefan Banach International Mathematical Center, to appear.

[Vigué, 1] Vigué, J.-P., Le groupe des automorphismes analytiques d'un domaine borné d'un espace de Banach complex. Application aux domaines bornés symétriques, Ann. Sci. Ec. Norm. Sup. (4) 9 (1976) 203-282.

[Vigué, 2] Vigué, J.-P., Automorphismes analytiques des produits continus de domaines bornés, Ann. Sci. Ec. Norm. Sup. (4) 11 (1978) 229-246.

[Vladimirov, 1] Vladimirov, V.S., Methods of the Theory of Functions of Many Complex Variables (The M.I.T. Press, Cambridge, Mass., 1966).

[Wright, 1] Wright, M.W., The behavior of the infinitesimal Kobayashi pseudometric in deformations and on algebraic manifolds of general type, Proceedings of Symposia in Pure Mathematics, vol.30, Amer. Math. Soc. Providence, R.I., 1977, Part 2, 129-134.

[Yosida, 1] Yosida, K., Functional Analysis, II ed. (Springer-Verlag, Berlin-Heidelberg-New York, 1968).

[Zorn, 1] Zorn, M.A., Gateaux differentiability and essential boundedness, Duke Math. J. 12 (1945) 579-583.

[Zorn, 2] Zorn, M.A., Characterization of analytic functions in Banach spaces, Ann. of Math. (2) 46 (1945) 585-593.

SUBJECT AND SYMBOLS INDEX